CLEAN ENERGY TECHNICIANS

PRACTICAL CAREER GUIDES
Series Editor: Kezia Endsley

Clean Energy Technicians, by Marcia Santore
Computer Game Development and Animation, by Tracy Brown Hamilton
Craft Artists, by Marcia Santore
Culinary Arts, by Tracy Brown Hamilton
Dental Assistants and Hygienists, by Kezia Endsley
Education Professionals, by Kezia Endsley
Fine Artists, by Marcia Santore
First Responders, by Kezia Endsley
Health and Fitness Professionals, by Kezia Endsley
Information Technology (IT) Professionals, by Erik Dafforn
Medical Office Professionals, by Marcia Santore
Multimedia and Graphic Designers, by Kezia Endsley
Nursing Professionals, by Kezia Endsley
Plumbers, by Marcia Santore
Skilled Trade Professionals, by Corbin Collins
Veterinarian Technicians and Assistants, by Kezia Endsley

CLEAN ENERGY TECHNICIANS
A Practical Career Guide

MARCIA SANTORE

ROWMAN & LITTLEFIELD
Lanham • Boulder • New York • London

Published by Rowman & Littlefield
An imprint of The Rowman & Littlefield Publishing Group, Inc.
4501 Forbes Boulevard, Suite 200, Lanham, Maryland 20706
www.rowman.com

6 Tinworth Street, London, SE11 5AL, United Kingdom

Copyright © 2020 by Marcia Santore

All rights reserved. No part of this book may be reproduced in any form or by any electronic or mechanical means, including information storage and retrieval systems, without written permission from the publisher, except by a reviewer who may quote passages in a review.

British Library Cataloguing in Publication Information Available

Library of Congress Cataloging-in-Publication Data

Names: Santore, Marcia, 1960– author.
Title: Clean energy technicians : a practical career guide / Marcia Santore.
Description: Lanham : Rowman & Littlefield, [2020] | Series: Practical career guides | Includes bibliographical references. | Summary: "Clean Energy Technicians: A Practical Career Guide includes interviews with professionals in a field that has proven to be a stable, lucrative, and growing profession."—Provided by publisher.
Identifiers: LCCN 2020011933 (print) | LCCN 2020011934 (ebook) | ISBN 9781538141687 (paperback ; alk. paper) | ISBN 9781538141694 (epub)
Subjects: LCSH: Clean energy industries—Vocational guidance.
Classification: LCC HD9502.5.C542 S26 2020 (print) | LCC HD9502.5.C542 (ebook) | DDC 621.042023—dc23
LC record available at https://lccn.loc.gov/2020011933
LC ebook record available at https://lccn.loc.gov/2020011934

Contents

Introduction: So You Want a Career as a Clean Energy Technician?	vii
1 Why Choose a Career as a Clean Energy Technician?	1
2 Forming a Career Plan	31
3 Pursuing the Education Path	57
4 Writing Your Résumé and Interviewing	85
Notes	115
Glossary	119
Resources	125
Bibliography	131
About the Author	137

Introduction

So You Want a Career as a Clean Energy Technician?

Welcome to a Career as a Clean Energy Technician!

Clean energy and renewable energy sources go hand in hand. Water, wind, solar, and geothermal all provide sources of abundant energy that are available directly from nature. Clean energy sources don't emit carbon into the atmosphere or radioactive waste into the environment. Renewable energy sources don't get used up. Together, these energy sources provide a sensible and cost-efficient alternative to the fossil fuels and nuclear power we have relied on in the past.

As a clean energy technician, you'll be at the forefront of making our world a better, cleaner, safer place. ©yangna/E+/Getty Images

We have the technology to provide energy throughout the United States and the world using clean and renewable energy sources, and we're developing newer and better technology all the time. We need people to develop, make, install, and maintain that technology to keep the energy flowing into our communities. That's where clean energy technicians come in.

Clean energy technicians (also called renewable energy technicians) work on all kinds of renewable energy technologies. As a clean energy technician, you'll specialize in a particular type of renewable energy technology. You'll be installing, maintaining, and managing the systems that create energy for our society. You'll be keeping up with the most current developments in the field. And you'll be helping the planet by promoting and providing renewable energy for the present and the future.

> The wealth of the nation is its air, water, soil, forests, minerals, rivers, lakes, oceans, scenic beauty, wildlife habitats and biodiversity . . . that's all there is. That's the whole economy. That's where all the economic activity and jobs come from. These biological systems are the sustaining wealth of the world.—Gaylord Nelson[1]

Careers in Clean Energy

There are many different kinds of careers in the field of clean or renewable energy—far more than can be covered in one book. Here's a quick list of just some of the available career options:

- Biofuels and Biodiesel Technology and Product Development Manager
- Biofuels Processing Technician
- Biofuels Production Manager
- Biomass Plant Technician
- Biomass Production Manager
- Civil Engineer
- Clean Car Engineer
- Customer Service Representative
- Electrical Engineer
- Electronics Technician
- Energy Auditor
- Energy Conservation Program Manager
- Energy Efficiency Analyst
- Energy Policy Analyst

- Energy Trader
- Environmental Engineer
- Environmental Engineering Technician
- Environmental Restoration Planner
- Environmental Science and Protection Technician
- Environmental Scientist
- Founder Energy Related Enterprise
- Geological Sample Test Technician
- Geothermal Installer
- Geothermal Production Manager
- Geothermal Technician
- Geoscientist
- Green Construction Manager
- Ground Crew
- Hydroelectric Plant Technician
- Hydroelectric Production Manager
- Hydro Mechanic Diver
- Instrument Control Technician
- Line Worker
- Machinist
- Mechanical Engineer
- Meter Technician
- Plant Electrician
- Power Distributor and Dispatcher
- Power Plant Operator
- Renewable Energy Consultant
- Renewable Energy Sales, Marketing, and Customer Service
- Renewable Energy Technician
- Resource Conservation Manager
- Solar Assembler
- Solar Designer
- Solar Energy Systems Engineer
- Solar Installer
- Solar Photovoltaic Technician
- Solar Project Manager
- Storage and Distribution Manager
- Substation Electrician
- Substation Operator
- Sustainability Professional
- Sustainable Builder
- System Operator
- Wind Energy Engineer
- Wind Energy Operations Manager
- Wind Energy Project Manager
- Wind Farm Site Manager
- Wind Turbine Technician

> Clean Jobs America estimated that more than 2.5 million people in the U.S. work in clean energy jobs. Energy efficiency accounted for the most jobs in that estimate, with renewable energy generation coming in as the second top employer. Solar energy workers make up a large portion of current career pathways in renewable energy, but don't think the possibilities stop there; more green energy jobs are coming.—Avery Phillips[2]

The Market Today

Clean energy is a growing field. According to the US Department of Energy's Office of Energy Efficiency & Renewable Energy, clean energy is "one of the fastest-growing, most innovative sectors of our economy," noting that one of the single fastest-growing occupations in the United States is wind turbine technician, which added twenty-five thousand new jobs in 2017 alone. Meanwhile, the office estimates "that one out of every 50 new jobs created nationally came from solar" adding, "The core of the industry and the technology's growth continues to be solar installers, who help set up and maintain the panels that are increasingly appearing on rooftops."[3]

There is a market for clean and renewable energy at all levels, from individual homes, to community and statewide efforts, to national policy. Areas with abundant sunshine turn to solar power, while those with the right wind speeds build wind farms. Hydropower can come from large installations like the Hoover Dam, which lies between the states of Nevada and Arizona, or from tiny pico hydro systems that generate less than 5 kilowatts. Your community might find that geothermal is the way to go. Or perhaps the best option is a combination of different technologies, like the hybrid wind and solar project built by Juhl Energies in western Minnesota.[4]

What Does This Book Cover?

In this book, you'll get a general overview of what four different kinds of clean energy technicians do and what to expect at different stages of your career.

STEP 1: WHY CHOOSE A CAREER AS A CLEAN ENERGY TECHNICIAN?

In the first chapter, you'll learn about certain clean energy technician jobs, what they are, what they do, what they earn, and what you can expect in these careers:

- Wind Turbine Technicians
- Solar Photovoltaic Installers
- Hydropower Technicians
- Geothermal Technicians

We'll also take a look at advancing in your career—from entry-level to experienced professional.

STEP 2: FORMING A CAREER PLAN

The second chapter is all about you and how you can plan your career as a clean energy technician. What do you need to know about yourself? What kind of clean energy career suits you best? How can you make your time in high school work for you? Where can you find more information?

STEP 3: PURSUING THE EDUCATION PATH

The third chapter shows you what kind of education you need for each career. You'll learn what to consider when choosing your educational opportunities, how admissions and financial aid work—and some things to watch out for.

STEP 4: WRITING YOUR RÉSUMÉ AND INTERVIEWING

In the fourth chapter, you'll learn about applying for and keeping a job as a clean energy technician, including important skills like writing a résumé and a cover letter, filling in an application form, and interviewing like a pro.

Where Do You Start?

There are so many directions you can go as a clean energy technician. Discover the path you want to take by turning the page and taking the first step.

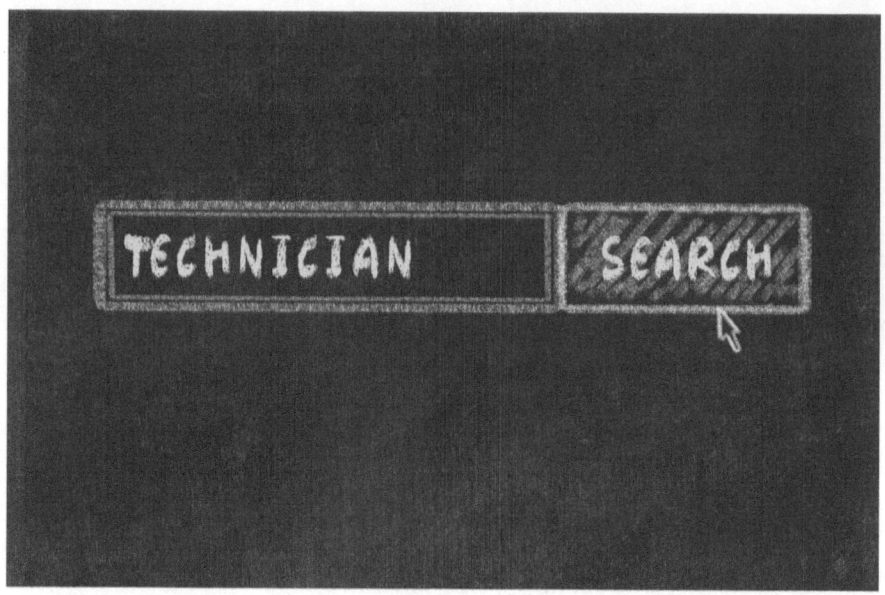

Employers will be searching for you when you're a qualified Clean Energy Technician! ©IUshakovsky/iStock/Getty Images Plus

1

Why Choose a Career as a Clean Energy Technician?

About Clean Energy

Clean (or renewable) energy is generally defined as energy (that is, electricity) that comes from renewable, zero-emissions sources. Clean energy sources like solar, wind, water, and geothermal renew themselves naturally and can't be used up. Clean energy does not emit carbon into the environment, so

As a clean energy technician, you are part of the solution. ©*Mintr/iStock/Getty Images Plus*

it does not add to the problems of global climate change. Clean energy doesn't release dangerous chemicals or radiation into the environment, either, so it is safer for people and other living things.

We're already experiencing the effects of our long-term dependence on oil and petroleum products on the planet's climate. While some people continue to maintain that global climate change is either (a) not happening or (b) not our fault, the scientific facts are clear. The problem is no longer one that may happen in the future—the problem is already here. Using clean and renewable energy is the only way to keep climate change from getting even worse.

Dependence on oil also creates unstable conditions between countries—especially between the ones that have oil and the ones that want it. Additionally, oil drilling on land and in the oceans has created numerous environmental disasters all around the world. Clean, renewable energy can be generated locally and regionally, keeping our energy production closer to home.

Clean, renewable energy is helping to make things better. Clean energy technicians are an important part of that. As Lora Shinn of the Natural Resources Defence Council notes,

> Renewable power is booming, as innovation brings down costs and starts to deliver on the promise of a clean energy future. American solar and wind generation are breaking records and being integrated into the national electricity grid without compromising reliability. . . .
>
> While renewable energy is often thought of as a new technology, harnessing nature's power has long been used for heating, transportation, lighting, and more. Wind has powered boats to sail the seas and windmills to grind grain. The sun has provided warmth during the day and helped kindle fires to last into the evening. But over the past 500 years or so, humans increasingly turned to cheaper, dirtier energy sources such as coal and fracked gas.
>
> Now that we have increasingly innovative and less-expensive ways to capture and retain wind and solar energy, renewables are becoming a more important power source, accounting for more than one-eighth of U.S. generation. The expansion in renewables is also happening at scales large and small, from rooftop solar panels on homes that can sell power back to the grid to giant offshore wind farms. Even some entire rural communities rely on renewable energy for heating and lighting.[1]

> **CLEAN ENERGY LEADERS—STATES TO WATCH**
>
> The Interstate Renewable Energy Council (IREC) announced its 2020 list of "those states that have demonstrated notable achievements and/or movement toward clean energy goals" on January 22, 2020.[2]
> Here's who made the list!
>
> - *New York*—Most Ambitious Clean Energy Goals, for bold clean energy legislation and ambitious energy storage goals
> - *Colorado*—Growing Clean Energy Leader, for tackling progressive regulatory reforms to the utility business model and laying a foundation for clean energy growth
> - *Massachusetts*—Most Charged for Storage, for policies and actions to boost energy storage deployment
> - *Virginia*—Nascent Market Poised for Clean Energy Growth, for addressing financial barriers to solar energy for low- to moderate-income households
>
> Keep an eye on these states!

What Is a Clean Energy Technician?

Today, renewable energy technologies provide about 20 percent of the electricity used in the United States. But that percentage is going to be growing, and clean energy technicians will be needed more and more to keep that clean, renewable energy available to the people who need it.

Clean energy technicians—also known as renewable energy technicians—are people with the skills and aptitude to ensure that the technologies associated with clean energy work at peak efficiency. Clean energy technicians usually specialize in a particular area, such as wind, solar, geothermal, or hydropower. Clean energy can be generated on a large scale for an entire state, country, or region, or it can be generated on a small scale for individual homes or as part of a "net metering" plan that allows individuals to sell power back to the grid. Clean and renewable energy sources are still sometimes called alternative energy (meaning instead of petroleum-based and nuclear sources of energy) but in the twenty-first century, clean and renewable energy is increasingly mainstream. It won't be "alternative" for long.

As a clean energy technician, you may find yourself working indoors or outdoors, in all kinds of weather, sometimes high in the air. You'll be using both your intellectual skills and your physical skills to keep equipment functioning and power flowing. This is considered a green career—that is, a career that is beneficial to the environment.

There are many different types of clean energy resources, and technicians are needed for all of them. In this book, we're going to take a closer look at technician careers in four types of clean energy:

- Wind Turbine Technicians
- Solar Photovoltaic Installers
- Hydroelectric Plant Technicians
- Geothermal Technicians

Clean energy technicians are working hard every day to make our world a better place.

What exactly is the renewable energy sector? . . . Renewable energy is defined as any energy that is naturally renewed. It is infinite; it does not run out. Renewable energy is also clean. Its use results in virtually no emissions or associated pollution. And as an added benefit, renewable energy can also be solely produced within the country. This creates self-sufficiency so that the country's energy needs don't have to be met through imports from other nations. The sector consists of any renewable energy company or organization that is committed to the research, development, advancement, and implementation of clean technologies and fuels.
—Trade-Schools.net[3]

Jobs in Clean Energy

Let's take a look at the different ways we can use clean and renewable resources to generate electricity, and at the technician jobs that are so important in making that energy possible.

> In this [electrical power generation] sector, USEER reports solar power supporting 242,343 jobs, coal supporting 86,202 jobs, and natural gas supporting 43,526 jobs [in 2018]. Renewable energy firms surveyed for the USEER report highlight that a substantial barrier to increasing employment is finding skilled labor to fill positions.— Environmental and Energy Study Institute (EESI)[4]

What Is Wind Power?

The kinetic power of the wind can be captured by wind turbines and turned into electricity. You've probably seen wind turbines—those tall white structures with three long blades. Sometimes you'll see just one, but often you'll see them in groups, called wind farms.

The American Wind Energy Association (AWEA) explains how wind turbines work this way:

> When the wind blows past a wind turbine, its blades capture the wind's kinetic energy and rotate, turning it into mechanical energy. This rotation

Wind turbine technicians at work. ©*aydinmutlu/E+/Getty Images*

turns an internal shaft connected to a gearbox, which increases the speed of rotation by a factor of 100. That spins a generator that produces electricity.[5]

The smallest scale for wind power is distributed wind. This refers to a small-scale turbine that generates less than 100 kilowatts. Distributed (or small) wind is used directly to create energy for a home, a small business, or a farm. Distributed wind turbines are usually off-grid—meaning they are stand-alone systems that are not connected to the local electric utility.

Wind turbines at the utility scale are larger and can generate from 100 kilowatts up to several megawatts. These are the kind of turbines found on wind farms; they send electricity to the grid and the local power company sends it out to users.

The largest turbines are found offshore. These huge turbines are built in large bodies of water (like the ocean) and generate much more power than land-based turbines.

In a single year, the U.S. wind resource potential could produce 364.9 quadrillion BTUs, the energy equivalent of all proven oil and natural gas reserves in the U.S. as estimated by the Energy Information Administration (EIA). A renewable resource, wind resource will not be depleted and will continue to provide energy year after year.—Christine Real de Azua, AWEA[6]

Wind power is plentiful all over the world and can never be used up. As one of the fastest growing renewable energy technologies, wind power has the potential to become one of the most important sources of clean energy. Between 2008 and 2018, the amount of electricity in the United States that was generated by wind power increased by 600 percent. The percentage of electricity provided to US communities by wind power is expected to grow considerably in the future.

WINDMILL OR WIND TURBINE—WHAT'S THE DIFFERENCE?

People often point to a wind turbine and call it a windmill. But are they really the same thing?

Both windmills and wind turbines are machines. Both have large blades that turn in the wind. And both generate energy—but not the same kind of energy.

There are all kinds of windmills. They have been used for generations to pump water, grind grain into flour, and for other jobs. But windmills generate mechanical energy, not electricity.

Wind turbines, on the other hand, are sophisticated modern technology. Wind turbines are designed to generate electricity from the kinetic (moving) power of the wind. Whether used one at a time or in large groups on a wind farm, wind turbines are an important part of the way we generate electricity now and in the future.

Windmills and wind turbines have similarities and important differences. ©bluejayphoto/ iStock/Getty Images Plus

Wind Turbine Technicians

Wind turbine technicians maintain wind turbines to ensure that they are operating safely and at peak efficiency. A wind turbine technician (or windtech) installs, maintains, inspects, operates, and repairs wind turbines. Windtechs diagnose and repair any problem that might shut the turbine down unexpectedly.

Some wind turbine technicians work on building new wind turbines, but most of the job duties relate to servicing existing turbines, especially the nacelles (the housing for the parts of the turbine that work together to generate electricity, like the generator, gearbox, drive train, and brake assembly). As a windtech, you might find yourself

- Climbing towers to inspect and repair equipment
- Collecting turbine data for testing/analysis
- Performing routine maintenance
- Testing components and systems
- Replacing old/malfunctioning components
- Servicing various systems (underground transmission, wind field substations, fiber optic sensing/control)

WORKING CONDITIONS

Wind turbine technicians need to be comfortable with small spaces as well as great heights. Work can be indoors but is primarily outdoors.

Wind turbines are often around 260 feet high, and windtechs carry more than forty-five pounds of equipment as they climb. Because of the risky nature of the job, wind turbine technicians generally work with at least one partner. According to the DOE's Office of Energy Efficiency & Renewable Energy,

> Technicians do most of their work in the nacelle, where sensitive electronics are housed. Since they are built compactly, technicians must be comfortable working in confined spaces. In addition, they also work on top of the nacelles, where they might have to replace instruments that measure wind speed and direction or work with large cranes. To do this, they are standing literally hundreds of feet in the air. To protect them, they wear fall protection, full body harnesses that are attached to the nacelle.[7]

WHY CHOOSE A CAREER AS A CLEAN ENERGY TECHNICIAN?

Wind turbine technicians need a good head for heights. ©*inakiantonana/E+/Getty Images*

What goes up must come down. Windtechs must know how to rappel down the side of the wind turbine or down a blade in order to reach the area that needs to be serviced.

WORK SCHEDULE

You can expect to work a standard eight-hour day and forty-hour week as a wind turbine technician, but you might be assigned to day, night, or weekend shifts. In addition, you may be called out beyond your regular work hours to deal with emergencies.

PAY

Wind turbine technicians earn an average of $54,370 annually, which is about $23.13 per hour. Entry-level windtechs with no previous experience would be likely to earn around $17.78 per hour, while the most senior wind tech with a lot of experience might make up to $36.66 per hour (more than $76,000 a year).[8]

PREPARATION

It takes a combination of education and training to qualify as a wind turbine technician.

Education

You can learn to be a wind turbine technician through certificate or associate degree programs offered by community or technical colleges after completing a high school diploma or getting a GED. These programs address the mechanical, electrical, hydraulic, computer, and other systems that go into a wind turbine, as well as the skills to troubleshoot and repair those systems. In addition, you'll most likely learn safety and rescue skills, along with first aid and CPR training.

Training

Training varies by employer, but most manufacturers typically provide one or two years of on-the-job training under the supervision of a wind turbine servicing contractor. You can also start training while you are still in your educational program.

License/Certification

There is no mandatory license or certification for wind turbine technicians. However, as the US Bureau of Labor Statistics (BLS) points out, "Professional certification can demonstrate a basic level of knowledge and competence. Some employers prefer to hire workers who are already certified in subjects such as workplace electrical safety, tower climbing, and self-rescue. There are many organizations who offer certifications in each of these subjects, and some certificate and degree programs include these certifications."[9]

SKILLS AND QUALITIES

In addition to the specific skills learned in a certificate or associate's degree program, wind turbine technicians need to have:

- Mechanical ability to be able to maintain and repair all the complex mechanical, hydraulic, braking, and electrical systems they'll be working with on a daily basis
- Physical strength and stamina to lift heavy equipment, climb, rappel, and then do it again
- Troubleshooting skills to figure out in the moment what the problem is and then repair it
- Detail orientation to make precise measurements, follow all safety procedures, complete tasks in order, and keep meticulous records
- Communication skills to communicate with other windtechs, supervisors, manufacturers, utility companies, and the public as needed
- Cool heads to stay calm when dealing with problems or facing taxing or difficult physical conditions to get the work done

THE FUTURE

The BLS predicts that the need for wind turbine technicians will grow by 57 percent by the year 2028—much more than most other jobs. The BLS adds, however, that "because it is a small occupation, the fast growth will result in only about 3,800 new jobs over the 10-year period."[10] Location also makes a difference—wind farms are more likely to be found in the Midwest, the Great Plains states, and near coastlines, making jobs in those areas more plentiful.

What Is Solar Power?

Solar power is energy that comes from the sun. Solar energy is what makes life on earth possible. It is the most abundant and most renewable energy source we have. We can also capture the power of the sun and convert it into electricity. The most common way to do this is by using solar photovoltaic (PV) or thermal capture. Solar PV panels are most common for smaller-scale projects, such

Solar photovoltaic installer inspecting solar panels. ©RuslanDashinsky/E+/Getty Images

as on homes or other small installations, while solar thermal capture is used for large-scale utilities.

More than 2 million US households are already using solar panels on their roofs. Across the country, solar farms use large numbers of solar panels to provide electricity to companies or utilities. Manufacturing solar panels creates a small amount of pollution (an ongoing challenge) but solar power itself is completely carbon-free and nonpolluting.[11]

> Solar panels convert solar energy into usable electricity through a process known as the photovoltaic effect. Incoming sunlight strikes a semiconductor material (typically silicon) and knocks electrons loose, setting them in motion and generating an electric current that can be captured with wiring. This current is known as direct current (DC) electricity and must be converted to alternating current (AC) electricity using a solar inverter. This conversion is necessary because the U.S. electric grid operates using AC electricity, as do most household electric appliances.—Jacob Marsh[12]

Solar Photovoltaic Installers

Solar photovoltaic installers assemble, install, and maintain solar PV systems. Sometimes these systems are on individual homes or businesses; sometimes they are in larger groupings called a solar farm. Beginning solar PV installers usually start with more basic tasks and work their way up to more complex work under the supervision of an experienced worker. Some states allow solar PV installers to connect solar panels to the grid, but others require an electrician to do that work.

WORKING CONDITIONS

Most PV installation work is done outdoors, often on rooftops of houses or small to mid-sized businesses, or other high places. Indoor work is usually in new buildings or inaccessible areas like attics and crawl spaces that are not climate controlled. Be prepared to be very hot or very cold!

If you work on a solar farm, you might work at ground level. It's likely you'll be part of a team that includes roofers and electricians.

There are some risks involved for the solar PV installer, such as falling off a roof or a ladder, or getting an electrical shock or a burn from malfunctioning or misused equipment. Anytime you work outside, you could be bit or stung by insects. If you work on roofs, you'll need to wear fall-protection equipment, along with more usual protective gear such as work boots and a hard hat.

WORK SCHEDULE

Most solar PV installers work a forty-hour week—sometimes more. The work schedule can be irregular because it depends so much on the weather. Production demands or the length of the contract can also affect which hours and how many hours you work in a given week.

PAY

On average, solar PV installers make about $42,680 a year. An entry-level worker might expect about $30,180 (depending on location), while a very experienced solar PV installer could make up to $63,580.[13]

PREPARATION

Education

You'll need a high school diploma or GED, plus courses from a community/technical college or trade school in relevant subjects, such as basic safety, PV system design, and installation techniques. Someone with prior construction experience, such as an electrician, might be able to take these classes online. The BLS predicts that jobs will be most plentiful for solar PV installers who have completed a community/technical college program.

Training

On-the-job training can last from a month to a year. This is where you'll develop your hands-on skills and learn even more about tools, PV system installation techniques, and especially safety. Job prospects are improved for those with on-the-job training like an apprenticeship or related experience such as journey-level electrician training. The companies that manufacture solar PV systems often provide specific training on their own systems.

License/Certification

You don't need a specific license to be a solar PV installer, but most employers will expect you to have a driver's license. There is no required certification, but being certified by an organization such as the Electronics Technicians Association, International (ETA International), the North American Board of Certified Energy Practitioners (NABCEP), and Roof Integrated Solar Energy (RISE) indicates that you have demonstrated that you are a competent solar PV installer.

SKILLS AND QUALITIES

In addition to the specific skills learned in a certificate or associate's degree program, solar PV installers need to have:

- Mechanical skills to work with complex electrical equipment and to build support structures for solar panels
- Physical strength and stamina to lift and carry heavy equipment, climb ladders, and crawl through tight spaces

- Detail orientation to follow complex and detailed instructions
- Communication skills to communicate with clients and coworkers
- Cool heads to stay calm and focused to be sure the job is done correctly and safely

THE FUTURE

Solar PV installers have a bright future (pun intended). The BLS predicts the career will grow by 63 percent by 2028, considerably faster than average for all occupations. The long-term outlook for the job depends in part on local and state incentive programs to help reduce the cost of installation for homeowners.

SOLAR OPPORTUNITIES FOR VETERANS

Veterans are being encouraged to enter the field of solar energy! Check out these programs if you or someone you know is a veteran of the United States Armed Forces. Here's what the Solar Foundation has to say:

> Veterans of the U.S. Armed Forces are outstanding candidates for careers in the solar industry. Military service provides the leadership abilities and technical skills that solar companies value highly. While some veterans begin with entry-level jobs and move up the ranks, others transition directly to advanced leadership roles within the rapidly growing solar workforce. . . .
>
> Solar Ready Vets Fellowship
>
> Service members from select military bases will be placed into 12 week on-the-job training fellowships with solar employers to facilitate their transition from active duty to civilian careers in the solar industry. This initiative will be part of the successful Hiring Our Heroes Corporate Fellowship program. Service members will come from select military bases in regions with high demand for solar workers.
>
> The fellowship program will be focused mainly on management and professional roles, such as technical sales, system design, supply chain logistics, project development, and operations, in addition to installation. Through placement with industry employers, service members will receive valuable on-the-job training, professional development, and career guidance.

> Solar Opportunities and Readiness (SOAR) Initiative
>
> The SOAR Initiative is leading the way to expand solar career opportunities for veteran populations. Through partnerships among solar companies, training providers, and workforce development networks in high-demand markets, this initiative connects veterans with solar credentialing and professional development opportunities.[14]
>
> Learn more at https://www.thesolarfoundation.org/solar-ready-vets/.

What Is Hydroelectric Power?

Hydroelectric power plants harness the power of moving water to generate electricity. Hydroelectric power can be produced at almost any scale, from massive hydropower plants that use huge dams to funnel water into giant turbines and generate more than 30 megawatts to small (micro or even pico) hydro plants that can be built and operated by a single person and generate up to 100 kilowatts. Other kinds of hydrokinetic technology harness electricity from the power of the ocean's tides and currents.

> **THREE WAYS WATER MAKES ELECTRICITY**
>
> - *Impoundment* is used by large hydropower systems to store river water in a reservoir. The release of the water spins turbines that generate electricity. How much water gets released depends on how much electricity is needed and the need to maintain a constant water level.
> - *Diversion* (also called run-of-river) channels some of the water in a river into a specific area to activate a turbine and generate electricity. Diversion might use a small dam or no dam at all.
> - *Pumped* storage is a method to save electricity from other sources by pumping water from a lower level to a higher level. When demand for electricity is high, water can be released back to the lower level, where it spins a turbine to generate electricity.

Large hydropower plants help provide electricity for entire regions. ©doranjclark/iStock/Getty Images Plus

Power can be harnessed even by very small, very simple hydro generators. ©yuliang11/iStock/Getty Images Plus

Hydroelectric Plant Technicians

Hydroelectric plant technicians (also called hydro technicians) work at power plants that generate electricity from water. They operate, monitor, and control power plant equipment, including turbines, pumps, valves, gates, fans, electric control boards, and battery banks. They make sure the equipment is operating correctly and make adjustments and repairs to ensure optimal performance.

WORKING CONDITIONS

Most hydroelectric plant technicians work at mid- to large-sized hydroelectric power plants. You might be working indoors, or you might be outdoors in all kinds of weather. You'll be working with large equipment that might be very high and/or rotating, so proper attention to safety as well as wearing safety equipment is very important. Many hydroelectric plants are located in remote areas, so expect to have a long commute to work as well as from the company to other work locations.

WORK SCHEDULE

Since power plants have to operate twenty-four hours a day, seven days a week, hydroelectric plant technicians might be assigned to any shift and should expect some evening and weekend hours and overtime assignments.

PAY

The average salary for a hydroelectric plant technician is about $57,100, which is equivalent to around $27.45 per hour. Entry-level jobs start at around $16.63 an hour ($34,580), while more senior positions earn closer to $39.30 per hour ($81,750 per year).[15]

PREPARATION

Education

An associate's degree in engineering technology is considered the minimum credential for a hydroelectric plant technician. In some cases, other training or work experience can substitute for the associate's degree, but most employers want to see that formal education on your résumé. Technological advances and increasing automation mean that opportunities will be best for those who also have learned about electrical engineering, information technology, or computer science.

Training

According to Dr. Anh Speer of Virginia Tech, "To work in this field it helps to have previous work-related skill, knowledge, or experience. Employees in these jobs are required anywhere from a few months to one year of working with experienced employees. A recognized apprenticeship program may be associated with these jobs. Some companies provide a paid hydro technician training program that prepares you to enter the higher level of hydroelectric plant technician once the program is completed."[16]

Some companies provide paid on-the-job training to be certain their hydro technicians know exactly what they need to know. For example, the Tennessee Valley Authority (TVA) offers a training program for those who meet certain educational and experience prerequisites, including two years of high school or college algebra. Those who qualify and enter the program receive a starting salary of $44,765 during the first of five periods, ending the program with a salary of $63,960 and qualification as a hydro technician.[17]

License/Certification

Requirements vary by state. Some states require hydroelectric plant technicians to be licensed as engineers or firefighters. Others require specific credentialing only for specific job functions. Be sure to know and understand the rules in the state where you plan to work.

SKILLS AND QUALITIES

To be successful as a hydroelectric plant technician, it's important to have certain skills and qualities in addition to your training. For instance, you'll need to have:

- Concentration skills to concentrate on important tasks over a long period of time without getting distracted

- Detail orientation to stay organized and thrive in a highly structured environment
- Mechanical skills to work with large, complex hydroelectrical equipment
- Dexterity as fine motor movements are as important as large ones
- Physical strength and stamina to lift and carry heavy equipment, climb ladders, and work in all kinds of conditions
- Problem-solving skills to find problems and fix them quickly and efficiently

THE FUTURE

Competition for hydroelectric plant technician jobs may be tighter than for other clean energy technician careers in the next ten years or so. While the market for hydroelectric plant technicians will grow about 2.5 percent by 2026, that figure is based on the belief that seventy-five hundred current technicians will be retiring. Newer, more efficient plants that make increasing use of automation could require fewer technicians to run. However, technicians with more advanced education and training will be needed to operate those more complex systems.[18]

What Is Geothermal Power?

Geothermal energy is created using the earth's own heat. Electricity is produced in geothermal power plants by accessing hot water from deep underground, in either liquid or vapor form. Hot water is pulled up one or two miles from below the surface of the earth, and the steam it makes moves a turbine to generate electricity. Later, the steam is cooled in a cooling tower and returned to the ground as cold water, where it will reheat and the process will repeat.

As of 2011, there were completed geothermal power plants operating in nine states: Alaska, California, Hawaii, Idaho, Nevada, New Mexico, Oregon, Utah, and Wyoming; Colorado, Louisiana, Mississippi, and Texas were developing geothermal projects. The largest of these is the Geysers, located in the Mayacamas Mountains north of San Francisco, California. The Geysers draws from natural steam field reservoirs and takes up forty-five square miles. It gen-

Geothermal power uses the earth's own heat to generate electricity. ©Lisa-Blue/E+/*Getty Images*

erates about 725 megawatts of electricity, which can power 725,000 homes (equal to a city the size of San Francisco).

Geothermal energy can also be used directly for heating (or even cooling) individual homes without first converting it to electricity. This direct form of geothermal energy can be drawn from relatively shallow ground. At about four feet underground, the temperature is a constant 55 degrees Fahrenheit. Geothermal heating systems use pipes buried at least that deep, which pump liquid to absorb the heat and then carry it into the building via a heat exchange device. When the weather is hot, the system can run in reverse—absorbing warm air from the building and returning it underground.

Using geothermal energy for heat is very efficient and very green. Almost no energy is wasted in the process, which is good for the environment and for the wallet.

Geothermal Technicians

Geothermal technicians work at both the larger power plant level and the smaller commercial/residential level.

In a power plant, geothermal technicians maintain, test, calibrate, operate, and repair all the advanced technology that is used to create electricity from hot water from deep underground. Modern power plants have both mechanical and computer equipment, all of which must operate at peak efficiency. So geothermal technicians in a power plant must be prepared to diagnose and repair problems as part of regular maintenance and on an emergency basis. The many components exposed to air and water can be subject to corrosion, so a geothermal technician may need to apply coatings to equipment or structures to mitigate that damage.

At the residential level, geothermal technicians install, test, and maintain the equipment that makes it possible to use this green energy source to heat and cool buildings. This includes calculating what the building's heating and cooling needs will be, planning where the various components of the system will go, digging and refilling ditches for heat pump loops, as well as installing and maintaining pipes and equipment at every point in the system. You'll need to know and understand the manufacturer's specs for the systems you use, help lay out the system, maintain safety at the job site, make sure the customer understands the system after it's installed, and respond to repair calls.

Geothermal heat pump coils go only four to six feet below the earth's surface. ©*BanksPhotos/iStock/ Getty Images Plus*

WORKING CONDITIONS

Geothermal technicians—whether working for a power plant or on residential buildings—should be prepared to be outdoors in all kinds of weather. Protective safety equipment like hard hats, safety shoes and glasses, gloves, and hearing protection are all required. As in any job of this nature, it's possible to be exposed to contaminants or to hazardous conditions, so it's important to pay attention. Traveling to job sites should also be expected.

Geothermal technician jobs at the residential level are available all over the country, since the ground temperature is pretty much the same everywhere. According to the EIA:

> Most of the geothermal power plants in the United States are in western states and Hawaii, where geothermal energy resources are close to the earth's surface. California generates the most electricity from geothermal energy. The Geysers dry steam reservoir in Northern California is the largest known dry steam field in the world and has been producing electricity since 1960.[19]

A large company will provide all the tools you need to do the job, as well as protective uniforms. A smaller company might expect you to provide your own hand tools and protective gear.

WORK SCHEDULE

Geothermal technicians usually work a standard forty-hour week, with the occasional need to be available for evening, weekend, or holiday work. Emergencies can arise in power plants or in people's homes—it's important to be able to respond quickly and get things running properly again.

PAY

The rate of pay for geothermal technicians varies depending on whether you work in a power plant or for a smaller residential geothermal heating company, as well as with education, experience, and location.

For instance, average pay for someone working for a heating, ventilation, and air conditioning (HVAC) business that works on homes, schools, and commercial buildings is $47,610 annually (about $22.89 an hour). An entry-

Geothermal power plants find earth's heat deep underground. ©*Burkay Dogan/iStock/Getty Images Plus*

level geothermal technician would make closer to $29,460 a year, while one with a lot of experience could make up to $76,230. In addition, geothermal technicians in the state of Alaska average about $61,550 annually. Of course, you should always consider the cost of living in a particular state when comparing average salaries in different parts of the country.

A geothermal technician or geothermal operator working in a power plant might average somewhere between $46,000 and $78,000 annually. However, there are not that many geothermal power plants in the United States, so these jobs are not as easy to come by.

PREPARATION

Education

A high school diploma or GED is required to be a geothermal technician. It helps to have taken classes in math, science, and electrical/electronics equipment installation and repair. Those intending to work on the residential installation side of geothermal would benefit from HVAC courses and certification. Certificates and associate's degrees are available through community or technical colleges. At a power plant, associate's or bachelor's degrees in electrical engineering can qualify you for higher salaries and advancement up the career ladder.

Training

Commercial/residential geothermal technicians usually train on the job under the supervision of a more experienced professional. Manufacturers of geothermal heating units often provide training on their specific systems. It's important to have a good understanding of math and science, as well as strong mechanical, technical, and computer skills. The North American Energy Reliability Corporation (NERC) is a source for training and continuing education.

License/Certification

NERC also provides certification, which is required for geothermal power plant operators if their position could affect the power grid.

SKILLS AND QUALITIES

Geothermal technicians and power plant operators need the right skills and qualities to be successful. For instance, you'll need to have:

- Mechanical and electrical skills, to work with large machinery as well as both hand and power tools
- Computer and technology skills, to use computers and devices for many purposes
- Math and science skills, to understand and calculate within systems
- Detail orientation, to ensure all installation and maintenance procedures are done correctly, as well as record collected data, and follow all safety rules and regulations
- Physical strength and stamina, to shovel, climb, carry heavy equipment, and work long hours in varying weather conditions

- Problem-solving skills, to find problems and figure out how to fix them
- Communication skills, to work with customers, clients, coworkers, and supervisors

THE FUTURE

HVAC mechanics and installers—including geothermal technician jobs on the residential side of the business—are projected to grow by 13 percent by 2028, which is faster than most occupations in the United States. As the BLS explains, "Technicians who specialize in new installation work may experience periods of unemployment when the level of new construction activity declines. Maintenance and repair work, however, usually remains relatively stable. Business owners and homeowners depend on their climate-control or refrigeration systems year round and must keep them in good working order, regardless of economic conditions."[20]

Summary

Clean energy technicians have so many options, and you've had a chance to learn about four of them. The work is interesting, mentally and physically challenging, important, and plentiful. As a clean energy technician, you'll be helping to provide the electricity that powers our homes and communities. And you'll be making the world a cleaner, safer, better place for all.

So where do you start? The next chapter is all about planning!

CHARLES HAUCK—SOLAR ENERGY

Charles Hauck is operations and maintenance coordinator for United Renewable Energy in Alpharetta, Georgia. United Renewable Energy develops, designs, builds, and maintains solar photovoltaic and energy storage systems for utilities, industrial and commercial companies, independent power producers, and electrical membership cooperatives. The O&M division is responsible for maintenance on

WHY CHOOSE A CAREER AS A CLEAN ENERGY TECHNICIAN?

completed projects, maintaining the sites and making sure everything is working properly and issues are resolved.

How did you decide to go into the field of solar energy?

I spent fifteen years in the sporting goods industry. I've always been involved in construction as well, and thought this would be a great opportunity. The solar industry is still in its teenage or young adult years. There's a lot of upside because of the need for energy. Every minute, more and more energy is needed. Customers are seeing the benefit of it. It seemed like a great field to get into when I wanted to start a new career.

Charles Hauck is O&M coordinator for United Renewable Energy. *Photo courtesy Charles Hauck*

What is a typical day on your job?

I manage a portfolio of customer-owned sites. Our company mainly works with utility-scale projects, so our customers are either utilities or are companies that generate energy to sell to the grid. Operations monitors the sites, using third-party software to see what's happening on the site without being there. Is the site communicating, how much sun or radiance is out there that day? How much is it producing that day? It also alerts you to issues that may have popped up. The day starts by monitoring these dashboards to be sure that everything is working as expected. If not, we contact the customer and say something's happening here. We get approval to go out on site, troubleshooting and problem solving whatever the issue is. There's scheduling, getting hold of technicians or subcontractors, all the back-end stuff that's required depending on what the issue is. Figuring out pricing, things you need to have the customer approve. Finally, creating a report for the customer about the resolution of the problem. Operating and maintaining those sites on a day-to-day basis. You can go weeks without having any issues, and then one day you have lots of sites go out at one time. That becomes a different level of reporting and problem solving!

What's the best part of your job?

The most satisfying part is if there is an issue, diving in and figuring it out. Being able to resolve the issue, resolve it quickly, and let the customer know so they can continue to reap the benefits of their investment. The customer gets paid when their solar site is producing energy, so we want to get these sites up and running as soon as possible. And that keeps the lights on at our house! You walk away with a real sense of accomplishment.

What's the most challenging part of your job?
Sometimes you think it's one thing that needs to be fixed, but it might be something else. Trying to figure out what the issue is before you send a technician out to the site, so they aren't sitting there tapping their feet while you're still trying to figure it out. That's where it can cost more to the customer, once someone is out there. And if it's something critical, making sure we have the people available who can get out there as quickly as possible to get it resolved. Another challenge is that problems always pop up unexpectedly, so you have to rearrange your schedule so that all the critical problems are taken care of.

What's the most surprising thing about your job?
The time required to manage each site. It depends on the size, but each one is their own little property with their own systems; they have grass, they have fences. Even with preventative or annual maintenance, if they discover a missing screw on a piece of equipment, it means time to find that specific part, get it delivered and installed. All the little five minutes here and ten minutes there, and the day is done.

How did (or didn't) your education prepare you for the job?
I came from a different industry prior to being in the solar industry, so my education per se didn't directly help with the job. But my work experience was probably the most important. Having key skills like attention to detail, troubleshooting, problem solving, time management, how to write a report clearly with detail, customer interaction—knowing how to speak to site owners, that there's proper interaction in a professional manner. Those types of skills are key. Since I've been here, I've also taken electrical classes and solar classes, so that I can understand even more about the industry.

Is the job what you expected?
I came into this industry with an open mind. I really didn't know what to expect. One thing I've learned in the electrical construction industry, safety is *so* key. Safety first is truly right. You're working with electricity. Electricity is the silent killer, the sleeping giant. It's always present—especially when the sun is shining. A lot of these sites are quiet, out in the country—you don't realize that it's utility-scale energy running through them. How safe and cautious you have to be always has to be at the top of your mind—where you stand, what you touch, are my tools rated for that equipment and functioning properly? It's those types of things that I wasn't expecting, but now

I have a heightened sensitivity to them. I'm aware of where I'm putting my hands, and if I have the right protective equipment on. Just leaning up against a post could close a circuit and cause severe issues. The first thing anyone on a site talks about is safety; the site manager goes over the safety rules, and if you're not rated or don't have the right equipment, you can't be there. You have to be safety minded to be in this industry.

What are the opportunities for advancement for a solar panel installer/technician?

On the O&M end, you can always grow your portfolio. There are customers who have sites built years ago who may not have the most up-to-date equipment or understand the maintenance and safety needs. In a company like ours, there's a lot of opportunity. On the sales side, there's going out and finding customers who are interested in building sites, and building relationships with them and bringing development projects to them. On the construction side, there are project managers who work directly with the customers when the sites are built. On the technical side, if someone has an electrical license or is an electrician, there's always need for that in this field. Technicians can go through the electrical training to build their skills. Outside of our business, there are also all the different products out there—the equipment that's required to build a solar site. If you don't have the product to build it, it can't be built! Someone who gets into a company like ours, there are opportunities—safety managers, quality assurance teams, there's a lot of opportunity to grow.

Where do you see the career of solar photovoltaic installer/technician going from here?

As solar has become more affordable and demand for solar and for energy in general continues to increase, there are many opportunities for a technician or installer. If you like to work with your hands and be out in the elements to build these sites, there's plenty of opportunity out there. We're also finding on the utility side, companies like to diversify—they want solar and wind and natural gas. All these different ways to produce energy. It's a great opportunity for someone right out of school to get involved in this career.

What is your advice for a young person considering this career?

An installer or technician should have the knowledge of how electricity works, like Electricity 101 or Solar 101, to understand an electrical circuit or a solar circuit. Wiring a house or being a commercial or residential electrician is slightly different from solar. Different voltages, more DC as well as AC. Understanding safety is always key. If you come aboard, you'd be apprenticed to an electrician or technician to learn hands-on about safety, how to wire something properly. After that, it depends on what you want to do, where you want to take it. Our company posts jobs online on Indeed or LinkedIn type sites. Doing research on the solar industry and seeing what's out there. Utility companies are always looking for someone to work in this sector. If you're willing to travel and work different places, there's that opportunity, too. It's not the typical nine-to-five job where you sit behind a desk all day. You can see the world—they're building solar sites all over.

2

Forming a Career Plan

What's So Important about Planning?

Some people know what they want to do from their earliest years and also seem to know what to do next to reach their goals. But if you're like the vast majority of people, you can see at least a few different options for your future, and you may not have much idea of how to make any of those goals a reality.

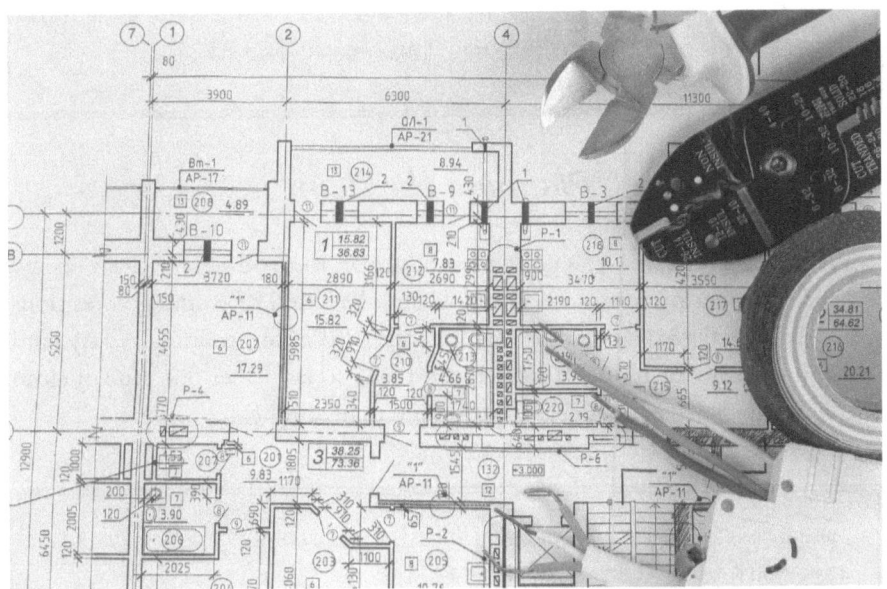

Plan your future as a clean energy techncian. ©Iurii Garmash/iStock/Getty Images Plus

Taking some time to consider the different career options out there as well as your own qualities—what you do best, what you like best, what you don't like at all—can help you narrow down those many options to just the ones that interest you most.

And it's not just important for your first job; it's important to plan ahead so your first job can lead to the next one and the next one, on up the career ladder. When you first get out of school, you may just be focused on your "launch"—finding a job that will pay for your daily, monthly, and yearly living expenses: an apartment, a car, and the things you like to do for fun. And that's fine for a while. But it won't be long before you realize that there's more to life than that. What about all those expenses you have that you didn't really think about earlier (like insurance of all kinds)? What about having a family and home of your own? What about vacations? What about retirement? The earlier you start planning for your future, the better prepared you will be for it when it suddenly becomes the present.

> Two types of choices seem to me to have been crucial to tipping their outcomes towards success or failure: long-term planning, and willingness to reconsider core values. On reflection, we can also recognize the crucial role of these same two choices for the outcomes of our individual lives.—Jared Diamond[1]

Planning the Plan

In the previous chapter, you got an introduction to the different clean energy technician career options—who does what, what kind of working conditions apply to each type of job, and a look at the education, training, salary, and opportunities for each one. What do you think so far? Does one type of clean energy technician career appeal to you more than another?

> Have a vision, a goal of what you want to do. Unless you're sure of where you want to go, you'll never get there.—Yogi Berra[2]

So What Goes into Your Plan?

In order to plan your plan, you first need to think about yourself. What do you like and dislike? What are you good at and not so good at? What feels like a comfortable fit for you? Next, you'll need to look at the different types of clean energy technician careers and look for the overlap between you and your interests and what the job is actually like.

Another important part of planning is to understand what you can do in advance to get ready to apply for the job you want. How do you become qualified to work as a clean energy technician? What kind of education will you need? What kind of training will you need? And how do you go about getting it?

In this chapter, you'll learn where to start and where to go from there. Let's start by making a few lists.

What Are You Like?

Every good career plan begins with you. A good place to start is by thinking about your own qualities. What are you like? Where do you feel comfortable and where do you feel uncomfortable? Ask yourself the questions in the "All about You" box and then think about how your answers match up with a career as a clean energy technician.

ALL ABOUT YOU

Personality Traits
- Are you introverted or extroverted?
- How do you react to stress—do you stay calm when others panic?
- Do you prefer people or technology and machinery?
- Are you better at making things or explaining things?
- How much money do you want to make—just enough, or all of it?
- What does the word success mean to you?

Interests
- Are you interested in how things work?
- Are you interested in solving problems?
- Are you interested in helping people?
- Are you interested in moving up a clear career ladder, or would you like to move around from one kind of job to another?

Likes and Dislikes
- Do you like to figure things out or to know ahead of time exactly what's coming up?
- Do you like working on your own or as part of a team?
- Do you like talking to people, or do you prefer minimal interaction?
- Do you have a head for heights?
- Are you comfortable in small, enclosed spaces?
- Do you like to be outdoors or indoors?
- Do you like to figure out problems and solve them?
- Can you take direction from a boss or teacher, or do you want to decide for yourself how to do things?
- Do you like things to be the same or to change a lot?

Strengths and Challenges
- What is something you have accomplished that you're proud of?
- Are you naturally good at school, or do you have to work harder at some subjects?
- Are you physically strong and active or not so much?
- Are you flexible and able to adapt to changes and new situations?
- Are you better at math or at English?
- Are you better at computers or doing things with your hands?
- What is your best trait (in your opinion)?
- What is your worst trait (in your opinion)?

Remember, this list is for only you. You're not trying to impress anybody or tell anyone what you think they want to hear. You're just talking to yourself. Be as honest as you can—tell yourself the truth, not what you think someone else would want the answer to be. Once you've got a good list about your own interests, strengths, challenges, and likes and dislikes, you'll be in a good position to know what kind of career you want.

> Don't sit back and wait for life to happen to you. Have a plan and take the needed steps to create what you want.—Steve Maraboli[3]

What Are the Jobs Like?

Now it's time to think about the characteristics of the different jobs you might do. Take a look at the questions in the "About the Job" box and consider the similarities and differences between different clean energy jobs.

ABOUT THE JOB

- What kind of work will you be doing?
- What kind of environment will you be working in?
- Will you have regular nine-to-five hours, or will you be working evenings, weekends, and overtime?
- What kind of community will you be living in—city, suburb, or small town?
- Will you be able to live where you want to? Or will you need to go where the job is?
- Will you work directly with customers?
- What will your coworkers be like?
- How much education will you need?
- Do you need certification?
- Is there room for advancement?
- What does the job pay?
- What kind of benefits will the job provide (if any)?
- Will you join a union?
- Is there room to change jobs and try different things?

Compare and Contrast

As you learned in chapter 1, clean energy technician jobs have a lot in common. The four types of clean energy careers that we've been looking at share some important qualities:

- Understanding and respecting electricity
- Installing, maintaining, and repairing skills
- Ability to use hand and power tools
- Respect for safety—following rules, paying attention, and wearing required safety gear
- Working in remote locations
- Working outdoors in all kinds of weather
- Testing equipment
- Troubleshooting problems
- Maintaining detailed records and writing detailed reports
- Working with equipment at a very large or very small scale
- Helping protect the earth while providing needed power for people

But they also have some important differences that could make the difference between choosing one kind of job over another. Compare your options and consider questions like:

- How much education do I need?
- What is the starting pay?
- Is there room for advancement?
- What do I need to know before I start?
- Is there on-the-job training or apprenticeship?
- Are the working conditions okay for me?
- Where are jobs located?
- What are the advantages of this kind of clean energy technician job?
- What are the hazards or disadvantages that come with this specific type of clean energy technician work?

LARGE COMPANY OR SMALL COMPANY?

What kind of company do you want to work for? How do you decide? There are opportunities for clean energy technicians at every scale.

Large companies and small companies have both similarities and differences. Whether those qualities are pros or cons depends on you and what you want out of your work experience. Let's look at a few differences.

Large Companies

- Structure: Larger companies tend to have more organization in their organization. There's probably a clear hierarchical structure as well as a clear career ladder.
- Your role: In a larger company, your job is likely to be well-defined. You'll know what is expected of you and where you should be focused.
- Benefits: Larger companies sometimes have better benefits (health insurance, dental plans, etc.) because having a large pool of employees puts them in a better bargaining position with insurance companies. You are more likely to find a range of insurance options to choose from.
- Salaries and bonuses: Larger companies may be able to offer higher salaries and annual or target-based bonuses because they have more money than smaller companies.
- Location: When you work for a larger company, be prepared to be sent where they need you. With an international renewable energy company, for instance, you could end up anywhere in the country or in the world!
- Training and educational benefits: Larger companies often have extensive training programs or pay for most or all of your continuing education credits.
- Other opportunities: A larger company may have more types of jobs, meaning that you can move "sideways" into a different type of work as well as moving up the ladder in your original career track.

Small Companies

- Structure: Smaller companies are often more flexible than larger ones. This means they can respond to changes in the market quickly. When a nimble company pivots, you can pivot with it.
- Your role: In a smaller company, your job may grow and change more easily than in a larger company.

- Benefits: Smaller companies may offer only one health insurance option, and may offer fewer benefits than a larger company. But they are in a better position to be flexible when you have something unexpected come up—and that will depend more on your boss's attitude than on the rules a larger company would have to apply.
- Salaries and bonuses: Smaller companies may not be able to offer the highest salaries or bonuses.
- Location: When you work for a smaller company, you are most likely to be working where the company is based. That means you have a better chance of staying in your own community, if that's an important factor for you.
- Training and educational benefits: Smaller companies don't usually run their own training programs, but they do offer apprenticeships with seasoned workers. Some small companies will pay for training and continuing education at nearby community or technical colleges and trade schools.
- Other opportunities: With a smaller company, you have a better opportunity to see how a business is run, to learn about the organizational and management side of the company, which is very useful if you might want to be a boss yourself one day.

Let's look at some quick stats about each of the clean energy technician jobs we've been considering, and see if we can answer some of those questions.

Wind Turbine Technicians

- Education: Certificate or associate's degree in wind turbine technology
- Entry-level starting pay: $19.67 per hour[4]
- Know before you start:
  - Rescue, safety, first aid, and CPR training
  - Electrical maintenance
  - Hydraulic maintenance
  - Braking systems
  - Mechanical systems, including blade inspection and maintenance
  - Computers and programmable logic control systems

- On-the-job training/apprenticeship: Typically at least twelve months of on-the-job training
- Working conditions: Outdoors, heights, confined spaces, remote locations, physically demanding, noisy, clinical assessment, and climb test required
- Locations: Wind farms exist all over the United States as well as in other countries. The states that generate the most power from wind are North Dakota, South Dakota, Minnesota, Iowa, Nebraska, Colorado, Kansas, Oklahoma, New Mexico, Texas, Maine, Vermont, Oregon, and Idaho.
- Projected job growth: 57 percent[5]

Solar Photovoltaic Installers

- Education: Solar installation coursework at community/technical college or trade school
- Entry-level starting pay: About $15 per hour[6]
- Know before you start:
    - Safety
    - Tools
    - PV system installation techniques
- On-the-job training/apprenticeship: One month to one year, depending on your prior knowledge; roofing or electrician apprentices and journey-level workers can do training modules specific to PV installation.
- Working conditions: Outdoors on the ground or on roofs, indoors in attics and crawl spaces
- Locations: Solar panels are installed throughout the United States. Jobs may include travel to near or distant job sites.
- Projected job growth: 63 percent[7]

Hydropower Technicians

- Education: Minimum of associate's degree in engineering technology; special certification or license required depending on the state and the facility

- Entry-level starting pay: $16.63 per hour
- Know before you start: Some previous work-related skill, knowledge, or experience helpful
- On-the-job training/apprenticeship: Multiyear training/apprenticeship programs provided by employers
- Working conditions: Indoors, outdoors, noisy, heights, cramped conditions, possible exposure to contaminants, and so on
- Locations: Large hydropower plants or smaller low-impact hydroelectric facilities exist all over the world, including forty-eight of the US states.
- Projected job growth: About 2.6 percent by 2026

Geothermal Technicians

- Education:
    - Residential work: At least HVAC courses; certificate or associate's degree in HVAC makes you more employable
    - Power plant: NERC certification (see the sidebar on NERC)
- Starting pay: About $14 per hour; higher for someone with more education and experience
- Know before you start: Mechanical, technical, electrical, and computer skills
- On-the-job training/apprenticeship: Supervised on-the-job training plus training on specific products from the manufacturer
- Working conditions: Indoors, outdoors, and underground
- Locations:
    - Residential work: Geothermal heat pumps can be installed anywhere.
    - Power plant: Most geothermal power plants are located in the western United States.
- Projected job growth:
    - Residential work: 7.6 percent grown between 2017 and 2019 noted in the United States[8]
    - Power plant: Unclear whether jobs will increase, decline, or stay about the same

ROOM FOR ADVANCEMENT?

Advancement as a clean energy technician depends on work experience, continued training and education, and (of course) the recommendations of your supervisors. You can work your way up from an entry-level technician to a mid- and then higher-level technician, supervisor, project manager, or more. If you want to reach the highest levels in the field, it is a good idea to go for a bachelor's degree in electrical engineering or a similar field.

For instance, as a windtech you might start your career as a technician, then become a qualified electrician, advancing to maintenance supervisor and then up to maintenance manager. Or you might go from technician to team leader to operations manager, ending up as a site manager.

As a solar PV installer, you could become a journey-level electrician and move up to lead electrician, and from there to project manager or construction superintendent. You could even own your own solar panel installation company.

As a hydropower technician or geothermal technician working for a large utility, you would typically advance to a high-level technician position before going on to be a project manager or supervisor. With a bachelor's degree in engineering, you could move up into an engineering position and from there possibly into management. As a geothermal technician working at the residential level, you might have the opportunity to move up into management and potentially run your own HVAC company in the future.

Finally, what does your gut say? Listen to the little voice in the back of your mind that says, "That's the one for me!" or "No way!" You already know what you'll like and what you'll be good at. A good fit is very important—more important than salary or many other considerations. So think it through with your head—but also listen to your heart.

Where to Learn More

There is so much information on the internet—about jobs and everything else—that it can be hard to separate the good data from the noise. So let's take a look at how to find good information about careers as a clean energy technician.

START WHERE YOU ARE

If you're in high school, you can start with your guidance counselor. A guidance counselor's job includes talking to you about what you would like to do after graduation and helping you plan and make decisions. If you're working on a GED at a community college or someplace similar, make an appointment at the career office or just walk in the door. Be sure to tell the counselor what you are interested in doing. *Speak up!* Remember, counselors can't read your mind—tell them what you're thinking so they can give you appropriate advice. And don't stop there.

TALK TO PROFESSIONALS IN THE FIELD

Are there clean energy companies in your area? Many communities these days have solar panel installation companies, HVAC companies that include geothermal, large or small wind farms, and even smaller power plants on rivers or dams that are used to generate electricity for the grid. There are also local and regional utility companies that purchase electricity from various sources and send it out through the grid for use in homes, schools, businesses, and so on.

Give them a call and see if someone will give you a tour of the plant, an informational interview about what it's like to work there, or general advice about the field. You might be able to arrange an internship or a summer job that will let you start building your skills right away.

> Our company posts jobs online on Indeed or LinkedIn type sites. Doing research on the solar industry and seeing what's out there. Utility companies are always looking for someone to work in this sector. If you're willing to travel and work different places, there's that opportunity, too. It's not the typical nine-to-five job where you sit behind a desk all day. You can see the world—they're building solar sites all over.—Charles Hauck (see interview in chapter 1)

LOOK IT UP

Do some research! Check out books from the library and visit websites for the businesses and utilities in your area, as well as those of larger clean energy companies. Some have sites all over the world. And don't forget to see the Resources section at the back of this book!

> **PLANNING FOR MORE THAN JOBS**
>
> Planning a clean energy future is bigger than planning individual careers. It has to take place on a local, regional, statewide, and national level. The state of California set a planning example in 2018 by setting a goal of 100 percent carbon-free electricity by 2045. That's an ambitious goal, but not an impossible one!
>
> The Union of Concerned Scientists (UCS) outlined ten key strategies that will "make the electricity grid more flexible, while reducing reliance on fossil fuels and prioritizing the needs of impacted communities—key components of the clean energy transition."[9] According to the UCS website, California should:
>
> 1. Use electricity as efficiently as possible and reduce demand at times of day when renewable supplies are least abundant.
> 2. Generate renewable electricity from a diverse mix of sources.
> 3. Plan for an orderly and equitable transition away from natural gas.
> 4. Use renewable generation technologies to provide grid reliability services.
> 5. Invest in energy storage at various timescales and locations.
> 6. Enable greater integration of western electricity markets.
> 7. Unlock the value of distributed energy resources.
> 8. Electrify cars, trucks, and buildings.
> 9. Shift electricity demand to better coincide with renewable electricity production.
> 10. Promote high-quality jobs and workforce development.
>
> That final list item is where you come in! Clean energy needs you on the team.

What Do Employers Want?

> We're also finding on the utility side, companies like to diversify—they want solar and wind and natural gas. All these different ways to produce energy.—Charles Hauck (see interview in chapter 1)

Many of the largest clean energy companies use a diversified approach—solar in one location and wind in another, for instance. Many are international in scope. EDP Renewables, as an example, is headquartered in Madrid, Spain. They operate wind and solar farms in North America, South America, and Europe. They work with landowners through long-term lease arrangements, producing electricity that can feed the grid in those locations. EDP's Hampton Solar Park near Charleston, South Carolina, can produce 20 megawatts (enough to power about three thousand homes) each year. Their Lone Star Wind Farm in Abilene, Texas, produces 400 megawatts of power each year—enough to power ninety-one thousand average homes.

Just like these companies, you can prepare yourself to do more than one kind of job if you plan ahead. For instance, if you pursue an associate's degree or certification in HVAC at a community or technical college, you'll be in a position to apply for jobs with solar PV installation companies, geothermal heat companies, or other clean energy HVAC positions. If you know about electricity, welding, construction, heating and cooling systems, or mechanics—any and all of these skills make you an appealing prospect to an employer.

WHAT ARE EMPLOYERS LOOKING FOR?

When companies want to hire new employees, they list job descriptions on job search websites. These are a fantastic resource long before you're ready to actually apply for a job. You can read real job descriptions for real jobs and see what qualifications and experience are needed for the kinds of job you're interested in. You can see what sort of tasks you'll be carrying out in different kinds of clean energy technician jobs. You'll also get a good idea of the range of salaries and benefits that go with different levels of experience.

Pay attention to the required qualifications, of course, but also pay attention to the desired qualifications—these are the ones you don't *have* to have, but if you have them, you'll have an edge over other potential applicants.

Here are a few sites to get you started:

www.monster.com
www.indeed.com
www.ziprecruiter.com
www.glassdoor.com
www.simplyhired.com

Making High School Count

How can you get the most out of your high school education if you want to be a clean energy technician?

- The most important thing you can do is work hard and get the best grades you can.
- Take math and science classes.
- Take classes where you can learn about electricity and electronics.
- Take classes where you can learn to use hand and/or power tools.
- If your school doesn't offer these courses, find someone else to teach you how to use tools—a parent, an adult friend, or a neighbor.
- Graduate! Most clean energy technician jobs require at least a high school diploma or GED.

Getting Experience

Even though much of your training for any type of clean energy technician career will happen on the job or in a specific certificate or degree program, it helps to have previous experience. A quick online search of openings for different kinds of clean energy technician jobs shows that employers are looking for at least some relevant experience.

Summer jobs, after-school jobs, or internships in any construction field can be useful previous experience. That includes not only electrical, but roofing, plumbing, building—anything where you use tools under the supervision of a professional. Mechanical experience is also helpful.

Don't be afraid to contact a local business and ask them what they're looking for. If you can use tools and have a driver's license, they may be willing to train you without other previous experience. It all depends on their needs at the moment. It never hurts to ask!

> **NERC CERTIFICATION**
>
> The North American Electric Reliability Corporation is an international not-for-profit regulatory authority that works to reduce risks to the reliability and security of the grid. NERC develops and enforces standards for bulk power systems in the continental United States, Canada, and the northern portion of Baja California, Mexico. NERC is the electric reliability organization (ERO) for North America, subject to oversight by the Federal Energy Regulatory Commission (FERC) and governmental authorities in Canada. NERC's jurisdiction includes users, owners, and operators of the bulk power system, which serves more than 400 million people.
>
> Through its system operator certification, NERC maintains the required credentials for more than six thousand system operators working in system control centers across North America who must meet certain important qualifications. As its website explains,
>
>> NERC's system operator certification exam tests specific knowledge of job skills and Reliability Standards. It also prepares operators to handle the bulk power system during normal and emergency operations. Certification is maintained by completing NERC-approved continuing education program courses and activities. These industry-accepted qualifications are set through internationally recognized processes and procedures for agencies that certify persons.[10]
>
> For more information about the system operator certification program, visit the NERC website at https://www.nerc.com/pa/Train/SysOpCert/, send an e-mail to NERC.System.Operator.Certification@nerc.net, or call 404-446-9759.

Build Your People Skills

The ability to get along with people is one of three most important factors of a successful career in any field. It's especially important for clean energy technicians! You'll be interacting with lots of people—clients, coworkers, supervisors, and others. The ability to communicate well—both verbally and in writing—is essential to getting a job, keeping a job, and doing that job well. There are many situations where clean energy technicians provide direct customer service, even entering people's homes to install or access equipment.

There are some basic people skills that anybody can learn that are useful for everyone, regardless of their career plans. Here are some of those important skills.

RELATING TO OTHER PEOPLE

This can be summed up as treating others the way you would want them to treat you. For instance:

- Try to see the other person's point of view (even if it's different from yours). Empathy and compassion go a long way.
- Be understanding and respectful toward other people.
- Be patient—nobody is perfect (including you).
- Pay attention and show genuine interest—everyone has something interesting about them. Take the time to find out what it is!

COMMUNICATION SKILLS

Good communication is essential in work and in life. Basic communication skills include:

- Active listening: It's important to pay attention to what the other person is actually saying and responding to it, not just planning what you're going to say next.
- Speaking: Learn to express yourself simply and clearly in spoken words.
- Writing: Learn to express yourself simply, clearly, and with sufficient detail in writing. Avoid a lot of extra words that can confuse the reader, but don't leave out anything important.
- Body language: Nonverbal communication is just as important as verbal communication. Pay attention to the messages you're sending with your face, gestures, and posture and pay attention to the nonverbal messages you're receiving from the people around you.

CHARACTER

Your character includes your personality and the choices you make based on your values and beliefs. Important character traits include:

- Honesty and trust: Honesty and trust form the basis of any relationship, whether personal or professional. This includes trusting others as well as being trustworthy yourself. When you and your coworkers know you can count on each other, you can accomplish anything. And when clients know they can trust you and the work you do for them, you will be set up for a lifetime of success.
- A sense of humor: Knowing when to use humor to lighten a situation is a great people skill. Used appropriately, humor can defuse tension and conflict. And be able and willing to laugh at yourself.
- Being supportive and helpful: Offer to do a little more than required or to help someone out when they need it. If someone is having a hard day, be respectful of their feelings.
- Flexibility: Be ready to adapt to changing situations, conditions, and workflow.
- Good judgement: Choose your own behavior—don't just go along with something if your gut says it's not a good idea.

When you get to chapter 4, you'll learn more about putting your people skills into action to get and keep the job you want.

Summary

Now you know how to plan your plan! Everything you need to know is easily available to you. You just need to put out your hand and pick it up.

Your plan will contain:

- Insight into who you are and what you're good at (as well as what you'd like to avoid)
- What you want out of your career as a clean energy technician
- What the different kinds of clean energy technician jobs are like
- What kind of education and experience you need to achieve your goals
- Ideas to improve or hone your people skills
- Don't miss the Resources section in this book! You'll find plenty of additional sources to help you figure out your plan.

The next chapter looks at what kind of educational options are available and answers the most important question: How do you pay for it? Don't worry! It's easier than you might think.

TIM LIPE—HYDRO TECHNICIAN III

Tim Lipe is a Hydro Technician III at the Tennessee Valley Authority Cherokee power plant at the Cherokee Dam on the Holston River in East Tennessee. He holds an associate's degree in electronics from Walters State Community College and a bachelor's degree in management from Tusculum University. Tim trained in TVA's thirty-month-long Hydro Technician Training Program, which covers all aspects of operation, maintenance, and safety. The TVA is a federally owned corporation that built the Cherokee Dam in 1940 to help meet the demand for energy at the beginning of World War II. In addition to providing power, the dam helps control flooding and the reservoir is used for recreational purposes like swimming, boating, and fishing.

Tim Lipe enjoys his work for the TVA at the Cherokee Dam power plant in East Tennessee. *Photo courtesy Tim Lipe and the Tennessee Valley Authority*

How did you decide to become a hydro technician?

There were a couple of different reasons. I liked that it was a clean, renewable energy source. I started out in fossil with TVA, so I've been through two training programs. I was there about eight years. When I had the opportunity to come over to hydro, I liked that it was renewable. But the other reason is that hydro was more of a stable job, since the fossil fuels aren't doing as well. In fossil, I did one job, but in hydro you're what's called multi-skilled, so I do a bunch of different things on a daily basis—mechanical, electrical, lots of things. When the opportunity for the hydro training came along, I was excited to try it.

What is a typical day on your job?

There's probably not a typical day. But some of the things we do on a daily basis are we come in and have a prework briefing that covers what needs to be done that day, what the dangers are that may be involved, what someone who worked on that job recently did. We have certain weekly inspections we do, like the powerhouse inspection that takes about a week to completely check everything. Sometimes there's an emergency, like a sump pump that's stopped working, and we have to deal with that. We might have equipment that's scheduled to be used but has to be taken out of service. The Tech IV usually gives out the work. For some jobs I work on my own, and sometimes we pair up. There're only four techs that work here at Cherokee: one Tech IV, three Tech IIIs, plus one laborer, so a total of five people. Some hydro plants have more, some less.

What's the best part of your job?

Just the fact that you're not doing the same thing over and over again. You may have inspections one day and emergent work the next day. If there's one thing I like over another, it's the troubleshooting part. You really have to dig into your knowledge and skills to figure it out. I like that better than anything. Also, we work in such small groups that you basically become a family. I know all the people who work here, their kids—you get really invested in working with the same person every day. It's a perk of the job that you might not think about at first.

What's the most challenging part of your job?

Not becoming complacent. In the program, you might have a thousand blueprints for this plant, and you're in the classroom eight hours a day studying these prints. When you're out of the program, you don't concentrate on that as much. You don't use everything every day. My biggest challenge is to study the print and keep up on all the technical information, so when a problem does arise, I know where to go to for this electrical contact, for instance, and remember exactly what it does.

What's the most surprising thing about your job?

When I came from fossil, I was an operator only. We didn't use multi-skills. In hydro, we have pipefitter, mechanic, electrical work. I was a little intimidated because I'm not an expert in any of those things. I was surprised at how easily I learned it. That's due to the training program. Don't be afraid of trying something you don't already know. I settled in really well after the training, and I can do just about everything we need to do around here. I got an interview and a job offer for the hydro job, and I wasn't sure I wanted to take it because I didn't know certain parts of the job.

But I learned them in the training program. It's really turned out better than I thought it would be.

How did (or didn't) your education prepare you for the job?
I have an associate's degree in electronics and a bachelor's degree in management. In order to qualify for the fossil program, you had to have a certain level of education. I did the Fossil SGPO (Student Generating Plant Operator) training—that was an eighteen-month training program. I was an operator in fossil for around eight years. The Hydrotech Training Program is fifteen months in the classroom and fifteen months on the job for thirty months total. I finished that about seven years ago. Having an electronics degree helped, but TVA's training program is so extensive, they really prepare you. Once I got out, all I had to do was learn this particular plant, where equipment was, what may be a little different about this plant compared to what I studied. I felt like I was really prepared when I got out. TVA has a well-regarded training program across the industry. I could probably go anywhere across the country and be qualified for a job in this field.

Is the job what you expected?
Yes, I would say for the most part. Except for how easy it was for me to adapt to the different crafts.

What's next? Where do you see yourself going from here?
I'm actually in the Tech IV training program right now, so I may move up to Tech IV. I haven't really decided yet. The Tech IV is like the foreman or supervisor over the Tech III, but I like both of those jobs. The TVA has other jobs you could do, but me personally, I think I'd stay in one of these two jobs.

What kinds of advancement opportunities would a hydro technician be likely to have?
Some of the natural progression has been for Tech IVs to move into plant manager jobs. Some Tech III and IVs have moved up to HVDC [high-voltage, direct-current] in Chattanooga. Regional support supervisor is another possibility.

Where do you see the career of hydro technician going from here?
In this area, I don't see them building any more dams, but with the push for clean energy, I see this being a very stable job. I can see how hydro techs could merge into solar energy work, because it's basically the same type of electrical background. Our biggest part is not even producing power, it's maintaining water levels, lake levels, maintaining the sluice gates, keeping the levels lower in the winter and

higher in the summer. Even if we didn't produce power, there would still be a need for the hydro tech.

What is your advice for a young person considering this career?
If you're thinking about doing this job, the courses you'd need to study would be all the math and science courses, because that would be really beneficial in the job. You need a degree or to be a journey-level machinist or pipefitter or electrician. Don't be intimidated. All the utility industry has such good training programs, you will learn every aspect of the job. And it pays well.

The Keystone Dam in Oklahoma. ©*Zsteves/iStock/Getty Images Plus*

MELISSA AHO—GEOTHERMAL HEATING

Melissa Aho runs Ultra Geothermal, Inc., in Barrington, New Hampshire, along with her husband, Darren Rice. Ultra Geothermal provides geothermal heating for residences along with other HVAC services. It's a relatively small company, with nine employees, including many geothermal technicians.

How did you decide to go into the field of geothermal energy?

Many years ago, I was the international training manager for the state of New Hampshire, and my biggest client was GT Solar. Renewable energy was something I was very enthralled with. My then-boyfriend (now husband) owned an HVAC company. He knew of geothermal and had done it with other companies. In 2008, we decided to turn the business into geothermal because of the crazy cost of oil.

Melissa Aho runs a geothermal heating company. *Photo courtesy Melissa Aho*

What is a typical day for the geothermal technicians at your company?

From an installation standpoint, the technicians are lined up with a job to go. All the jobs we do are air-sourced, so our men go to a job site and put ducting in. Occasionally we do radiant floors. And we always do our own loop fields, so they're looping our ground source into homes. It's the way geothermal runs. Ground-source heat pumps create the heat exchange that brings heat from the ground into your home. That's the most efficient way in New England. The ground is always 50 degrees, so our equipment is in the home and brings in heat from under the ground. This can be from a vertically drilled borehole, four or five hundred feet in the ground, or a horizontal loop about six feet in the ground, stretching out across the entire yard. We do just new construction and existing homes. It takes about a week to do a normal-size house.

The service side is not so easy. The service technician installs mini splits for those who want them. They help do the geothermal start-ups for new construction. They get called out for no-heat service calls, take care of emergency service, help out with everything else—sheet metal install, low-voltage electrical wiring, installing refrigerants, central air conditioning system service from other companies.

What's the best part of your job?

I've always run every facet of the job. One thing that sets our company apart is that all our customers can talk to us directly. I can integrate and help out wherever needed. I love to do the accounting! I do our website design and home shows. My favorite part is interaction with customers. My husband also deals directly with the job sites and goes out to show the technicians what they're supposed to do. I also like to put on presentations and classes about geothermal at schools and colleges. Being a woman in a male-dominated field, I'm the only woman who shows up for HVAC trainings! I'm also a green realtor, so I focus on selling homes that have geothermal or solar systems.

What's the most challenging part of your job?

One of the issues is the lack of knowledgeable troubleshooters on the manufacturer's end. The techs are frustrated by the lack of trained people who can walk them through a problem. It doesn't happen very often, but there aren't enough people coming up in this field. Our technicians are mostly in their forties and fifties. That's a problem with every trade there is. That's part of why I try to promote this as much as possible. This is a job a robot cannot do! It's hands-on, and it's stressful on your body. The biggest challenge is not having enough people to help.

What's the most surprising thing about your job?

It was surprising to find myself in this industry. I was just going to help out for a year and somehow, here I am sixteen years later. On a business level, the most surprising and perplexing thing is that if you dig in and take the opportunities to network with the right people, you can be so successful. I was a single mom with no money or resources, and now we have a multimillion-dollar company. There are microloans available from SBDC [Small Business Development Center], the SBA [Small Business Administration]. I was surprised that I could actually achieve what I did.

How did (or didn't) your education prepare you for the job?

I went to school for fashion design and merchandising. I didn't expect to go into construction. I had to do what I could afford. I came from nothing—my parents were very hard working, my dad was a construction manager. I went to the University of New Hampshire and paid for it myself, which I couldn't afford. I changed to a school in Massachusetts, changed directions, and tried fashion design and merchandising. I became a paralegal. Those are the degrees that help me in my job today. Then I was hired at the Southern New Hampshire University international program. I got my bachelor's degree in business from SNHU. It taught me all my accounting principles and human resources (HR) stuff. You don't have to lock yourself into what you're studying. You need to network—if you don't, you're doing yourself a disservice. If you put yourself in front of the right people, you get hired from the way you present yourself and the people you meet.

Is the job what you expected?
My expectations were to just help grow a green energy business. The first week I stepped into the job, the oil prices were skyrocketing. You didn't know what you were going to be paying at the gas pumps. Everyone suddenly turned to geothermal. I had to learn about HR, writing an employee handbook, figuring everything out. I was flying by the seat of my pants. Now I'm the most organized person you could ever meet. My expectations now are very high—for myself and my employees. It took a while! I advise young people not to set too many expectations, because that can be limiting. Look at the most experienced people in the field you want to be in and hang around with them and listen.

What's next? Where do you see yourself going from here?
I'm kind of at a turning point in my life. I don't know what the next step will be. I do a lot of different things. It's great to have reached a point where I can choose what I'm going to do.

Where do you see the career of geothermal technician going from here?
We need to find more and more of them. Every year more geothermal systems are coming online. We hire commercial refrigerant technicians rather than HVAC technicians, because we can teach them the HVAC stuff. I don't see the world of the geothermal technician ever going away. It's good to have a gas license, too. We have only nine employees, and our people are cross-trained to do multiple things. Being flexible works out well for our technicians.

What is your advice for a young person considering this career?
It's not necessary to get a bachelor's degree to go into this kind of work. One underutilized resource is the technical and community colleges. They can specialize in HVAC. If people don't have a lot of funds, it's much more affordable. If kids are confused about what they want to do, if they go to a community or technical college for a year and get a certificate in something, you can cut out all the debt and get right to work. My company offers reimbursement for education. I didn't get my bachelor's degree until I was an adult.

Make sure that you can get a job doing what you want to do. Try out whatever you think you want to go to school for to make sure the degree you want to get actually leads to the kind of work you want to do. If you want to do HVAC, general contracting—interview for a job and find out what they're actually looking for. Make sure the path you're on is going to get you there.

I've been asked to do school and college presentations and to speak at graduations. What I tell these kids is that your paths in life keep forking off depending on your interests. And you can take the fork in the road that you didn't think you were going to go down.

3

Pursuing the Education Path

You're probably hearing a lot about the high cost of college tuition these days, along with questions about whether everybody needs to even go to college. You have looked over the clean energy technician jobs in this book and seen that there are some where you can begin with just a high school diploma.

But if you look closer, you'll see that no matter which clean energy technician career you're interested in, it helps to have more education under your belt after high school. The more education you have, the more employers want to hire you. You won't need as much on-the-job training, so you'll be more productive sooner. And you'll be in a better position to be promoted later on.

The right education helps you move forward quickly as a clean energy technician. ©*SolStock/E+/ Getty Images*

Pursuing some higher education can be cost effective if you do a little research ahead of time. It also helps to understand how financial aid works (see the section "What's It Going to Cost You?" and the sidebar "Not All Financial Aid Is Created Equal," later in this chapter).

> The beautiful thing about learning is nobody can take it away from you.—B.B. King[1]

What Do You Need?

A basic background, such as an associate's degree in electronics, is a good starting point for any clean energy technician job. Some community or technical colleges offer associate's degrees in renewable energy that cover the science of electricity, building construction, safety, general education courses, and major coursework in a particular area of clean energy. This kind of degree usually includes hands-on experience with equipment. In other cases, you might find an associate's degree or certificate program that is focused on one specific area.

Let's begin by looking at associate's degree programs for different types of clean energy technicians. Associate's degrees are offered through community or technical colleges. Online programs exist for these degrees, but they can't offer the hands-on training that in-person courses do.

Associate's Degrees

Associate's degrees will have a certain number of classroom hours devoted to general education subjects such as math, English, history, psychology, communications, and other useful subjects. Don't assume you only need the technical classes! General education courses ("gen eds") help put your technical courses into context and prepare you for the nontechnical aspects of your job, as well as for interacting with others on and off the job, and to move on to a bachelor's degree, if you decide to do that.

Be aware that an associate's degree is equivalent to the first two years of a bachelor's degree. So if you decide you want to do a bachelor's degree later on, you're already halfway there with your associate's.

WIND TURBINE TECHNICIANS

According to the BLS Office of Energy Efficiency, wind turbine technicians usually "learn their trade at technical schools or community colleges, where they can earn an associate's degree or a certificate."[2] An associate's degree typically takes about two years.

While different schools will set their own individual courses of study, a windtech associate's degree will most likely include these technical courses in addition to the gen ed courses:

- AC and DC Electricity and Circuits
- Automation
- Basic and Advanced Safety
- Basic, Intermediate, and Advanced Hydraulics
- Computer Logic and Systems
- Electrical Systems and Maintenance
- Introduction to Physics
- Introduction to Psychology
- Introduction to Solar Technology
- Introduction to Wind Systems
- Machine Wiring and Safety
- Mechanical Systems and Maintenance
- Renewable and Sustainable Energy
- Servo System Camming and Registration
- Technical Math
- Technical Reporting
- Turbine Maintenance
- Wind Tech Rescue
- Wind Tech Tools

A good windtech degree or certificate program will often have functional wind turbines available for students to work on to get hands-on experience and a sense of what the equipment is really like.

SOLAR PHOTOVOLTAIC INSTALLERS

A high school diploma or GED is the minimum educational requirement for solar PV installers. Additional coursework in PV systems design and installation as well as safety is available from community colleges and trade schools. Solar PV installers with associate's degrees relating to electricity, electronics, and engineering are in a better position to be hired and advance more quickly. Those with electrician licenses and/or bachelor's degrees in electrical engineering can move up into management positions with their companies.

Coursework for an associate's degree in solar PV is likely to include some combination of technical courses like these:

- Basic, Intermediate, and Advanced Energy Efficiency
- Digital Drafting
- Electrical Inspection and Codes
- Electrical Power Generation and Control Circuits
- Environmental Biology
- Environmental Ethics
- Environmental Politics and Policy
- Environmental Psychology
- Environmental Science
- Geographic Information Systems
- Introduction to Electrical Technology
- Introduction to Solar Photovoltaic Systems
- Photovoltaic Theory and Installation Techniques
- Print Reading for Construction
- Technical Calculations

HYDROPOWER TECHNICIANS

An associate's degree in engineering technology can be a good choice if you want to be a hydropower technician. To be ready to advance up the career ladder, a bachelor's degree in engineering technology, business management, electrical engineering, or a similar major is a good idea and a good investment.

An engineering technology associate's degree would add courses like these to the gen ed program:

- AC and DC Circuit Analysis
- Analytic Geometry and Calculus
- Basic and Advanced Amplifiers
- Chemistry
- Computer Programming Languages
- Digital Circuits
- General Physics
- Introduction to Electricity
- Principals of Clean Energy
- Energy Efficiency
- Environmental Technology
- Power Technology
- Materials Science
- Conventional Energy Technologies

GEOTHERMAL TECHNICIANS

Geothermal technicians can choose between an associate's degree in clean energy technologies or one in heating, ventilation, and air conditioning technology. Those working in a residential setting will often be doing geothermal work in the context of other heating and cooling systems and may prefer the HVAC degree. Geothermal technicians intending to work in geothermal power plants or for a utility company would probably benefit from the broader approach of a clean energy associate's degree.

An associate's degree program for a geothermal technician would be likely to include technical courses such as:

- Air Pressure Tests and Analysis
- Compressor Installation and Maintenance
- Electric Heating Systems
- Energy Efficiency
- Energy Management
- Environmental Technology
- Fundamentals of Power Production, Transmission, and Distribution
- Furnace Ignition Systems
- Geothermal Heating Technologies

- HVAC Troubleshooting Skills
- Introduction to Electricity
- LEED Core Concepts
- LEED Green Building Design and Construction
- Principals of Clean Energy
- Smart Grid Technologies
- Temperature Control

You learn concepts and the hands-on stuff in college. ©SDI Productions/E+/*Getty Images*

Bachelor's Degrees

In most cases, you can begin working as a clean energy technician without a bachelor's degree. But there are many advantages to having one!

Whether you're just starting out, have finished an associate's degree, or have been working as a clean energy technician for a while, you might want to consider a bachelor's degree. Having a bachelor's degree means that you have put in the effort to learn even more about your field and are better prepared to work at a higher level. Earning a bachelor's degree is not just about the subject matter in your courses. It includes developing critical thinking skills, analytical skills, and problem-solving skills, as well as giving you a broader view of the world and everything in it.

There are many bachelor's degrees available, depending on where you want to go. Do you want to go deeper into the engineering and technology side of the business? Do you want to move up into managing or owning a renewable energy business? Are you interested in environmental and sustainability policy?

If you're interested in any of these areas, look for a bachelor of arts or bachelor of science program. Different colleges and universities use different titles for similar programs, so here's a list of what you might find:

- Business Administration in Sustainability and Green Energy
- Business Management
- Electrical Engineering
- Energy and Sustainability Policy
- Energy Commerce
- Energy Land Management
- Energy Policy and Management
- Energy Resource Management and Development
- Environmental and Sustainability Studies
- Environmental Science
- Industrial Technology
- Mechanical Engineering
- Renewable Energy
- Renewable Energy Engineering
- Sustainable Energy Science
- Sustainability Science
- Sustainability Studies
- Sustainable Business
- Sustainable Development
- Sustainable Energy Management
- Sustainable Environmental Design
- Technology Management
- Wind Energy

Learning in the classroom gets you ready to work and learn more on the job. ©*Huntstock/iStock/Getty Images Plus*

Choosing the Right Program for You

There are many kinds of schools that offer the kinds of programs you'll need to pursue your career as a clean energy technician. So what do you need to consider?

> Be sure you aren't paying too much! Watch out for education scams that are just out to sell you expensive student loans. If you work through your local community or technical college and follow up on all the available scholarships for people going into the trades, you shouldn't need to take out a lot of student loans.

ADMISSIONS REQUIREMENTS

Admissions requirements are those things you need to have in place in order to apply to and enroll in the college of your choice. When you're considering a two-year or four-year college program, be sure you go to the admissions page of the school's website and find out what the requirements are. Here's what's most likely—but again, each individual college is a little bit different, so don't skip the websites!

Community College: Certificate or Associate's Degree Program

- High-school diploma or GED
- No SAT or ACT exams ("open admissions")
- Possibly a placement test to help you choose appropriate coursework
- Residency requirements may apply

Technical College: Certificate or Associate's Degree Program

- Possible age requirements (at least sixteen, seventeen, or eighteen)
- High school diploma or GED
- Placement test
- Different programs may or may not require an entrance exam
- Acceptable entrance exam scores often allow you to skip the placement test
- Residency requirements may apply

Four-Year College or University: Bachelor's Degree

Most four-year colleges and universities do not have open admissions, so the application requirements are more extensive. While community colleges admit everyone who meets the requirements, four-year colleges may turn down applicants if they don't have room in a particular program or if the applicant doesn't show enough academic promise.

Students applying to a four-year college or university program straight out of high school will most likely be expected to provide:

- High school diploma or GED, or high school transcripts if you haven't graduated yet
- Usually an SAT or ACT score (although a few four-year colleges have stopped asking for these)

- Sometimes one or more SAT subject tests
- Essay(s) and sometimes short answers to prompts
- Recommendations (two or three letters)
- Application fee
- Common App form or similar
- Proof of residency (for state colleges and universities)
- Résumé

Transfer students are those who already have an associate's degree from a community or technical college or who have completed college-level coursework at another school and want to have those credits counted toward their bachelor's degree. Some of these requirements may be waived for applicants with associate's degrees or enough transfer credits.

ACADEMIC ENVIRONMENT

Does the school offer the majors or certificate programs you want? Does it have the right level of degree program? What percentage of classes are taught by professors and what percentage are taught by adjunct instructors? Are adjunct instructors working professionals in the field? Does the school offer internships, cooperative education programs, or help to find apprenticeships? Does the coursework line up with state standards and codes?

FINANCIAL AID OPTIONS

Financial aid is something you have to look at carefully. Some employers will pay the cost of coursework taken by their apprentices or employees. If you get a deal like that, you won't need to worry about financial aid. But not every employer can or will make that offer. Some can offer reimbursement for only one or two courses a year, and some don't provide an educational benefit at all.

Don't worry if you need financial aid. There are many scholarships available, especially at the community college level, for technicians and trades workers. Does your college provide access to scholarships, grants, work-study jobs, or other opportunities? How much does the cost of school play a role in your decision?

SUPPORT SERVICES

Support services include things like academic counseling, career counseling, health and wellness, residence services, the financial aid office, information technology support, commuter services, and services for students who are disabled, or who have families, or who are lesbian, gay, bisexual or transgender. Some schools also have religious services, such as a chaplain. Before you choose a school, look through the website and be sure it provides the services you will need.

CLUBS/ACTIVITIES/SOCIAL LIFE

Most colleges have clubs and other social activities on campus, whether the student population is mostly residents or mostly commuters. Look for clubs related to the major you're interested in as well as clubs and activities that meet your other interests. College campuses have all kinds of things going on all the time, for students and for the local community—concerts, comics, plays, open mic nights, game nights, art shows, and lots of other things. Don't miss out!

SPECIALIZED PROGRAMS

Does the school or program you're looking at have any programs that meet your specialized needs? For instance, some institutions have programs specifically for veterans. Some address learning disabilities or mental health issues. If you might benefit from a specialized program like these, be sure the school you attend can meet that need.

HOUSING OPTIONS

What kind of housing options do you want and need? Does the college provide dorms? How many students will share a room? Are there on-campus apartments? Is there help with finding off-campus housing like apartments or rooms for rent? Some community and technical colleges mostly serve students who live off-campus and commute for classes. Be sure you have an affordable place to live!

TRANSPORTATION

If you live off campus, how will you get to school? Is there a bus system—campus or municipal? Is there a ride-share program? Could you ride a bicycle? Will you need to have access to a car? If the campus is large, is there an on-campus shuttle bus service that can get you around quickly? Is there enough student parking?

STUDENT BODY

What's the makeup of the student body? What's the ratio of males to females? Is there enough diversity? Are most of the students residents or commuters? Part time or full time? Who will you meet?

College is a great place to meet and get to know other people who share your interests. It's also a great place to meet and get to know people who are very different from you. On a college campus, you'll encounter people from small towns and large cities; of all different ethnic backgrounds, genders, and ages; studying or teaching many different topics. Be sure you take advantage of the opportunity to discover more kinds of people!

THE RIGHT FIT

One of the most important characteristics of a college program is finding the right fit. What does that mean? It means finding the school that not only offers the program you want, but also feels right. Many students have no idea what they're looking for in a school until they walk onto the campus for a visit. Suddenly, they say to themselves, "This is the one!"

While you're evaluating a particular institution's offerings with your conscious mind, your unconscious mind is also at work, gathering information about all kinds of things at lightning speed. When it tells your conscious mind what it's decided, we call that a gut reaction. Pay attention to your gut reactions! There's good information in there.

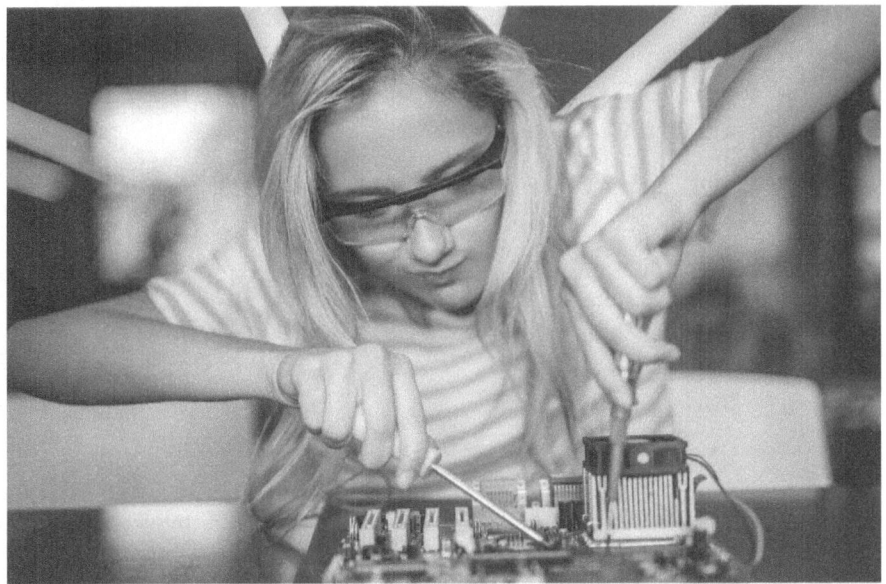

When you find the school for you, you'll know you've got the right fit. ©skynesher/E+/Getty Images

Beyond Graduation

Your degree is a great starting point, but you will need to keep up with licensing, certification, and continuing education in order to succeed as a clean energy technician.

CERTIFICATION

Requirements for certification also vary by state and by employer. Certification is not the same as licensing. In some cases, completing a certification program from an accredited trade school or community/technical college is all the certification you need.

Voluntary certification from a reliable organization like the North American Board of Certified Energy Practitioners is increasingly expected. It shows employers that you are a highly skilled and qualified candidate. It can make the difference between getting *a* job and getting *the* job.

According to the NABCEP website,[3] board certifications increase your marketability, validate your knowledge, and enhance your reputation, credibility, and consumer confidence. The organization offers the following board certification options:

- PV Installation Professional
- PV Design Specialist
- PV Installer Specialist
- PV Commissioning and Maintenance Specialist
- PV Technical Sales
- PV System Inspector
- Solar Heating Installer
- Solar Heating System Inspector

NABCEP also offers the Associate Credential, available by exam, which "recognizes individuals who have demonstrated knowledge of the fundamental principles of the application, design, installation, and operation of Photovoltaic, Solar Heating or Small Wind energy systems."[4] Associate credentialing is a first step toward board certification. It's a way for students, entry-level renewable energy workers, and workers in fields that don't offer professional certification to demonstrate their knowledge and skills.

The NABCEP has three ways to earn the Associate Credential:

- Education Pathway: Completing an NABCEP-approved training course in photovoltaics, solar heating, or small wind and passing the exam
- Experience Pathway: Documentation for at least six months of full-time work with photovoltaics, solar heating, or small wind technologies, plus passing the exam
- Entry-Level Conversion Pathway: NABCEP Entry Level Achievement Award for photovoltaics or solar heating, plus a small conversion fee; if your award is more than three years old, you'll need to show at least twelve hours of continuing education

LICENSING

Are there licensing requirements for clean energy technicians? It depends on the type of energy and on the state you live in. In general, any type of construction contractor needs an appropriate license from the state. If you own your own company, you'll need to be sure you have this paperwork in place before working on any projects.

In many states, separate licensing is not required for clean energy technicians. Of those that do, the most common requirement is a journey-level or master electrician license. (This requirement may include reciprocity agreements with other states.)

More states are starting to license solar installation as a separate contractor's license. Those doing residential geothermal installations usually have to comply with their state's license and certification requirements for HVAC contractors. Here's a quick sample of what some states require. It's not a complete list, so be sure you understand the requirements of your state.

- *Alaska*: Residential or nonresidential contractor's license is required to install solar thermal or PV panels, plus a Certificate of Fitness for various trades including electricians and those working with hazardous substances or boilers.
- *California*: The C-46 solar contractor license covers solar PV and thermal installations and maintenance. There are speciality licenses for active solar water and space heating systems, solar pool heating systems, and PV systems.
- *Florida*: A certified solar contractor license or an electrical contractor license is required for solar PV and thermal work. There are a number of subcategories for different types of solar installations and maintenance.
- *Oregon*: The limited renewable energy technician (LRT) license indicates someone is a specialty electrician trained to install renewable energy systems (solar electric, wind turbines, micro-hydro, and fuel cells) up to 25 kilowatts AC. Like other Oregon electricians, LRTs are regulated through the Building Codes Division and the Electrical and Elevator Board.

CONTINUING EDUCATION

Continuing education is always a good idea. It is also usually required to maintain licensing or certification. This generally consists of taking certain courses in your field from an accredited institution, such as a community/technical college, university, or trade school.

NABCEP Board Certifications require continuing education credits for recertification or certificate renewal, chosen from a catalog of more than two hundred possible courses.[5]

Depending on your location, you may be able to take the continuing education courses you need at a nearby community or technical college or even online. Your employer will be the best source of information as to which continuing education options you should use.

What's It Going to Cost You?

Costs can vary quite a lot depending on the field you want to go into and the program you choose. And there are a lot of other factors that affect the cost of your postsecondary education. Are you going to a two-year school or a four-year school? Is it public or private? How much financial aid are you eligible for in terms of scholarships or grants? How much will you be expected to borrow in student loans?

Yes, you can afford college! ©gustavofrazao/iStock/Getty Images Plus

Table 3.1 is based on information available on the College Board website and represents the state of things for the 2018–2019 academic year. (It's worth noting that these numbers represent a 2.5 percent increase over 2017–2018 costs before adjusting for inflation[6]). Costs shown are for one year.

Table 3.1. Annual Costs—Undergraduate College/University

2018–2019	Public Two-Year, In-District [AU]	Public Four-Year, In-State	Public Four-Year, Out-of-State	Private Nonprofit Four-Year
Tuition and Fees	$3,660	$10,230	$26,290	$35,830
Room and Board	$8,660	$11,140	$12,290	$12,680
Combined	$12,320	$37,430	$48,510	$48,501

That's a lot of money! However, these are averages. And note the difference in cost between a year at a two-year community or technical college and a year at a four-year college or university. In general, tuition and other costs for college tend to go up about 3 percent every year, so take that into consideration when planning for the year that you'll be going to school. You'll need to look closely at the costs of the schools you're considering—they could be quite different from the figures shown here.

There are all kinds of ways to get those costs down!

FINANCIAL AID

It is worth your while to put some time and effort into finding out what financial aid you qualify for. Reach out to the financial aid office at the school you want to attend. They can tell you a lot about what you may be able to work out.

Financial aid can come from many sources. The kind of awards you're eligible for depend on a lot of things, such as:

- Academic performance in high school
- Financial need
- Program/field of study
- Type of college

Follow up on financial aid for affordable higher education. ©designer491/iStock/Getty Images Plus

NOT ALL FINANCIAL AID IS CREATED EQUAL

Educational institutions tend to define financial aid as any scholarship, grant, loan, or paid employment that assists students to pay their college expenses. Notice that financial aid includes both *money you have to pay back* and *money you don't have to pay back*. That's a big difference!

Do Not Have to Be Repaid
- Scholarships
- Grants
- Work-study

Have to Be Repaid *with Interest*
- Federal government loans
- Private loans
- Institutional loans

TUITION BENEFITS

Before you apply for scholarships or loans, first find out if you are eligible for money for classes from outside sources.

Employer Tuition Benefit

Some employers are willing to pay some or all of the cost of coursework for their employees or apprentices. This isn't something every employer can afford, but if you are trying to choose among several employment options, a course tuition benefit can be an important factor in helping you decide. Even if the benefit is only a small amount, every bit helps.

Union Tuition Benefit

Are you in a union? As with employers, benefits vary from union to union and state to state. But unions provide important benefits to their members, and education is often one of them.

SCHOLARSHIPS

Scholarships are financial awards that are usually offered on the basis of academic merit, membership in Scouting or some other organization, or for going into a particular field—especially in an area like clean energy, where many more workers are needed. Scholarships can also be available to students who have certain characteristics, such as athletes, or who are underrepresented in a particular field or major, such as women or members of a minority group. Some scholarships go toward tuition; others are for specific things like textbooks and school supplies.

Scholarships usually pay a portion of tuition. It is very rare to receive a full-tuition scholarship, but it does happen. Scholarships do not have to be paid back. Scholarships can be local, regional, statewide, or national in scope.

There are also scholarships specifically for community college students, including those who want to transfer to a bachelor's degree program later on or those who are studying a particular subject. Some are offered by professional associations, some by nonprofit organizations, and some by the community colleges themselves.

Be sure to check with your high school guidance counselor as well as the school you're planning to attend. Some high schools have scholarships that go to graduating seniors who are planning to pursue a particular career, and many are restricted to community and technical colleges.

To learn more about scholarships, check out www.gocollege.com/financial-aid/scholarships/types/.

GRANTS

Grants are similar to scholarships. Most tuition grants are awarded based on financial need, but some are restricted to students in particular sports, academic fields, or demographic groups, or those with special talents. Grants do not have to be paid back.

Some grants come through federal or state agencies, such as the Pell Grant, SMART Grants, and Federal Supplemental Education Opportunity Grant. You'll need to fill out the Free Application for Federal Student Aid (FAFSA) form to apply for these grants. Learn more at https://studentaid.ed.gov/types/grants-scholarships.

Grants can also come from private organizations or from the college itself. For instance, some private colleges or universities have enough financial resources that they can "meet 100 percent of proven financial need." That doesn't mean a free ride, but it usually means some grant money to cover the gap between what the financial aid office believes you can afford and the amount covered by scholarships and federal loans. (More on federal loans below.)

WORK-STUDY

The federal work-study program provides money for undergraduate and graduate students to earn money through part-time jobs. Work-study is a need-based program, so you'll need to find out if you are eligible for it. Some students are not eligible at first but become eligible later in their college career. Most jobs are on campus and some relate to your field, but others—like working in the library—are more general.

Some colleges and universities don't participate in the work-study program, so check with the financial aid office to see if it's available and if you're eligible for it. It's good to apply early to have a better chance of getting the job you want most.

Since work-study is earned money (you do a job and get paid for it), this money does not need to be paid back. To learn more, check out https://studentaid.ed.gov/sa/types/work-study.

Loans

There is almost always a gap between the full cost of tuition and the amount of money you receive in scholarships and grants. That gap is filled by student loans. Student loans have to be repaid. Interest varies depending on the type of loan. Be sure you understand how much interest you will be charged, when the interest starts to accumulate, and when you must start paying the loan back. Usually, repayment starts when you graduate or after a six-month grace period.

FEDERAL LOANS

Federal student loans are issued by the US government. They have lower interest rates and better repayment terms than other loans, and you don't need anyone to cosign. If the loan is subsidized, the federal government pays the interest until you graduate. If it's unsubsidized, interest starts to accrue as soon as you accept the loan. That can amount to a very big difference in how much you pay for your education by the time the loan is paid off.

The most common federal student loan is the low-interest federal Stafford Loan, which is available to both undergraduate and graduate students. Depending on household income, a student's Stafford Loan might be subsidized or unsubsidized. (Another popular student loan, the federal Perkins Loan, is no longer available.)

Most schools will require you to fill out the Free Application for Federal Student Aid when you apply for financial aid. Note that it doesn't say "free student aid"; it says "free application." That means it does not cost anything to apply for federal student aid. You may get offers to submit the FAFSA for you for a fee—this is a scam. Don't do it.

PRIVATE LOANS

Chances are, if you are attending a four-year bachelor's degree program, federal student loans will not completely fill the gap between your tuition bill and any scholarships or grants you receive. Private student loans are issued by a bank or other financial institution. Rates of interest are generally higher than for federal loans, so be careful not to borrow more than you need. Eligibility criteria for private loans are based on credit history (both yours and your cosigner's).

Don't just take the first loan you find. Do some research and compare interest rates and terms. Is the interest rate variable or fixed? Is there a cap on the variable interest? Is the company reputable? What are the repayment requirements?

INSTITUTIONAL LOANS

Many educational institutions make their own loans, using funds provided by donors such as alumni, corporations, and foundations, as well as from repayments made by prior college loan borrowers. Every college has its own rules, terms, eligibility, and interest rates. Interest may be lower than private student loans, and the deferment option may be better as well.

Learn more about all kinds of financial aid through the College Board website at http://bigfuture.collegeboard.org/pay-for-college.

FINANCIAL AID TIPS

- Some colleges and universities offer tuition discounts to encourage students to attend, so tuition costs may be lower than they first appear.
- Apply for financial aid during your senior year of high school. The sooner you apply, the better your chances. Check out https://studentaid.ed.gov to see how to get started.
- Compare offers from different schools—one school may be able to match or improve on another school's financial aid offer.
- Keep your grades up—a good GPA helps a lot when it comes to merit scholarships and grants.
- You have to reapply for financial aid every year, so you'll be filling out that FAFSA form again!
- Look for ways that loans might be deferred or forgiven—service commitment programs are a way to use service to pay back loans.

While You're in College

While you're enrolled in college, of course you will take all the courses required by your program. But don't stop there. In addition to your technical courses, be sure to learn:

- Business management
- Technical writing
- Math
- Psychology

and enjoy all your general education courses.

Working While You Learn

Hands-on learning is an essential part of any technician job. While you're learning what you need to know to work as a clean energy technician, consider how to get that hands-on training at the same time.

If you already have a job and are also taking classes, you're golden—especially if your employer is paying all or part of your tuition! If not, look into one of these programs.

COOPERATIVE EDUCATION PROGRAMS

Cooperative education (co-op) programs are a structured way to alternate classroom instruction with on-the-job experience. There are co-op programs for all kinds of jobs; they are especially common in engineering programs. Co-op programs are run by the educational institution in partnership with several employers. Students usually alternate semesters in school with semesters at work.

A co-op program is not an internship (see below for more on those). Students in co-op jobs typically work forty hours per week during their work semesters and are paid a regular salary. Participating in a co-op program means it will take longer to graduate, but you'll come out of school with a lot of legitimate work experience.

Be sure the college you attend is truly committed to its co-op program. Some schools are deeply committed to co-ops as integral to education, but others treat it more like an add-on program. Also, the company you co-op with is not obliged to hire you at the end of the program, but can still be an excellent source of good references for you in your job search.

APPRENTICESHIPS

Most states have apprenticeship programs that allow you to contract with a licensed tradesperson to learn the job as an apprentice. If you're planning to be a clean energy technician, you might be able to arrange to apprentice with someone in the clean energy field you want to learn. Or you could apprentice under a master electrician and learn that trade, which will help you with any clean energy job you pursue.

Each state has its own requirements regarding what sort of contract or articulation agreement needs to be signed, how many hours/years are required, and where to submit all that information so you get credit for it. Some states require apprenticeship licenses, while others don't. If your state requires an apprenticeship license, you will need to show that you have a sponsor (employer) or approved program.

Apprentices are considered employees. As an apprentice:

- You will work full-time—that is, a forty-hour work week that will probably include evening and weekend shifts and emergencies.
- You will earn between $15 and $29 an hour, depending on your employer, location, and union status.
- You will receive the same benefits as other employees (vacation and sick leave, medical and other insurance, retirement plan, etc.).
- Some larger companies will reimburse the cost of coursework.
- Some companies will cover the cost of things like tools, work boots, and a company cell phone.

The company's expectations will be along these lines:

- You are qualified to be an apprentice according to your state's guidelines.
- You are age eighteen or older.

- You have earned a high school diploma or GED.
- You hold a valid driver's license with a good driving record.
- You can pass a drug test and a background check.
- You have good communication skills.
- You are a team player.
- You have the ability to problem solve and prioritize.
- You have flexibility in terms of available hours and the work you're willing to attempt and/or do.
- You have a good attitude.
- You are physically able to do the job.
- You are willing to commit to about five years in the apprenticeship.

PART-TIME JOBS

You probably won't have a hard time finding a part-time job in the clean energy field. Call a solar PV installation company or a geothermal HVAC company in your area and ask if they're hiring any part-time help. Or call an electrician or general contractor. You'll be surprised how in demand you're going to be. You'll find yourself up on ladders and rooftops as well as in crawl spaces and attics, but that's what you can expect as a clean energy technician once you're qualified. You won't get the complicated work that a clean energy technician needs to do, but you'll get good field experience on job sites and the chance to learn from experienced people—and you can earn some money at the same time.

INTERNSHIPS

Internships are another way to gain work experience while you're in school. They are last on this list for the simple reason that you don't get paid to do an internship. Internships are offered by employers and usually last one semester or one summer. You might work part-time or full-time, but you might not actually be doing relevant work while you're there.

There are paid internships in some fields, but they aren't common. Unless you happen to find a clean energy technician internship that also pays you or is with a company that you really, really want to work for, you should probably try the other options first.

Summary

Education? You've got this. You've considered the various clean energy technician jobs and what you need to know to work in each one of them. You've considered different kinds of educational settings to see which one would be the best for the future you're planning for yourself. You've learned all about financial aid and options for on-the-job learning. You're ready to start getting yourself ready to enroll!

Next, let's take a look at what happens after you complete your education and training. What goes into getting a job as a clean energy technician? And what do you need to know to be successful in your chosen career?

ALAN SMITH—HYDRO TECHNICIAN IV

Alan Smith is a Hydro Technician IV at the Tennessee Valley Authority Pickwick Hydro Plant on the Tennessee River in Southwest Tennessee. After serving in the Marine Corps and working on jet engines, Alan joined TVA, where he got his training in operation, maintenance, and safety in the Hydro Technician Training program. A federally owned corporation, TVA built the Pickwick Landing Dam in 1938. The plant has six generating units with a summer net dependable capacity of 247 megawatts. Pickwick Reservoir is popular for boating, water skiing, and sport fishing.

Alan Smith is in charge of an eleven-person crew at Pickwick Hydro Plant. *Photo courtesy Alan Smith and the Tennessee Valley Authority*

How did you decide to become a hydro technician?

I was a jet engine mechanic with Delta Airlines when 9/11 happened. My father worked for TVA; my grandfather had previously worked for TVA. My father suggested I come work here. I qualified for the veteran's preference program, and I got started with TVA as a laborer, doing janitorial work. That was March of 2004. After six months, my supervisor asked me if I wanted to go into the hydro tech training program. I started the training in September of 2004.

What is a typical day on your job?
We come in in the morning and have a safety meeting about the work that's going to be going on that day, who's going to do what task, and divide up the work. Then we go out to the work site. You do that task and then come back for another task. At the end of the day, we do a review to cover what you did for the next person who's going to be doing that job. I oversee the crew in the control room here at Pickwick. I'm usually the one assigning those tasks and checking the reports at the end of the day. The guys can go into our management system and put in their notes for the day on what they did. That's when we discuss that, and figure out what they're going to need, and get the materials on order for them.

What's the best part of your job?
I enjoy the smaller groups. We have eleven people here at Pickwick. It's a family environment. Everyone knows everyone's kids, who's playing ball, when our kids are playing ball against each other. And that it's multifaceted—the ability to be able to work on multiple different things, from looking at numbers on a report or troubleshooting why a generator won't start. Never the same thing two days in a row—that's for sure!

What's the most challenging part of your job?
The size of the crew and the variance of the work! Having such a small crew can be challenging. You have to be an expert in multiple facets of the job. You have to rely on the team and the diversity of the team to get the work done.

What's the most surprising thing about your job?
The level of technical components that continue to be put into place, and how TVA Hydro has developed since it was first built, from operators to computer programs. How technology continues to advance and play a part in my everyday job.

What was your education and training before and after you started at TVA?
I was a high school graduate. I had a year of college. I joined the Marine Corps. I did six years in the Marine Corps as a jet engine mechanic. I obtained my FAA air frames and power plants license before leaving the Marine Corps. I went to work for a FedEx feeder for ten months while my wife finished grad school, and then I got on with Delta Airlines in Atlanta. I worked there for about three years and went to different training schools for engines. I did take a few additional college courses

along the way—spotty units here and there. At TVA, I started the Hydro Tech Training Program in September 2004. It was a thirty-month program. I did fifteen months of eight-hour days in the classroom. And then I did fifteen months on-the-job training with more weekly assignments here.

How did this combination of education and training prepare you for the job?

Really well. I'll be biased and say I think the Marine Corps prepares you for life in general, as far as how you deal with challenges. Starting a jet turbine and a hydro turbine are more similar than most people would think. There's a lot of numbers involved in getting something spinning. It's really quite similar.

Is the job what you expected?

Yes. It's challenging and that's what I expected. I'm challenged on a daily basis. Even on the routine days you seem to find something that you didn't know or hadn't seen before. It's challenging and rewarding.

What's next? Where do you see yourself going from here?

I still feel like I have a lot to learn! I've been doing this job ten to twelve years now, but I still learn every day. From laborer to technician to lead technician at the plant—at some point management might possibly be in my future if the timing was right and everything worked out correct.

Where do you see the career of hydro technician going from here?

They built a lot of these hydro plants and some are very old. There's a lot of money being spent on clean energy. Hydro is clean energy; the fuel is free. I think it's a safe bet for a long-term career with a lot of different options in the United States and in the world, for that matter.

What is your advice for a young person considering this career?

I think you'd need to be committed to whatever you're going to do. Oh—and understand that hydro sites are usually in rural areas—if you're a city type, it might not be the best fit for you! But just to commit yourself to whatever you do, go full in, never look back, apply yourself, and give your all to whatever you do. Before I came here, I didn't necessarily know there was such a thing as a hydro tech job. It's really a very good job with good benefits. It's something more kids should pursue.

4

Writing Your Résumé and Interviewing

Putting It All Together and Getting the Job

You've done your planning and chosen a career as a clean energy technician. You've done your research and determined which kind of clean energy technician you want to be. You've prepared yourself with your education and training, and developed your practical skills as well as your knowledge of the job. And you know that being a clean energy technician is a great career with lots of opportunity.

A good résumé can get you that all-important interview. ©SDI Productions/*E+/Getty Images*

So how do you turn all that preparation into an actual job? This chapter takes a look at the other skills you need to get not just *a* job but *the* job. Those are the business skills that everyone—clean energy technicians included—need to have to be successful.

Writing Your Résumé

Everybody needs a résumé. Even if the job you're applying for requires you to fill in an online application form, your résumé will be where you've collected all the information the company wants in one place.

WHAT IS A RÉSUMÉ?

A résumé is a simple way to list everything you've done to prepare you for the job you're applying for. It includes sections for different things: your education, your training, your previous experience. It may also include any honors you've earned or special things you've done or been a part of.

Your résumé sums up your education, training, and experience. ©*iStock/Getty Images Plus*

You will submit your résumé (along with a cover letter) whenever you apply for a job. You may also want to upload your résumé to a few of the many résumé sites on the internet. A résumé should be one to two pages long.

For a clean energy technician's résumé, you want to be sure to include where you got your degree or training, if you did an apprenticeship or cooperative education experience (and where), what skills you have (especially the ones you do best), any specialized certifications you hold, and your employment history.

TYPES OF RÉSUMÉS

There are three basic formats for a résumé: reverse chronological, functional, and combined.

Compare the three sample résumés and you'll see that you may want to use a different type of résumé at different times or different stages in your career. Once you've got a lot of experience and want to move up to management, for instance, you'll want to highlight the specific skills you've acquired that qualify you for that role.

> The chronological style is the most popular with recruiters and hiring managers because it shows a clear time line of your employment. If you have a relatively steady work history and several relevant positions, the chronological format is usually your best choice.—Jeff Butterfield[1]

Reverse Chronological Résumé

The reverse chronological format is the most traditional type of résumé. A reverse chronological résumé is written with the most current information first, going backward to list the oldest information last. This type of résumé works for everyone, whether you're a student, an entry-level clean energy technician, or an experienced worker.

The usual layout for a reverse chronological résumé is pretty simple.

- Name and contact information at the top
- Education, starting with most recent first (apprenticeship, college, high school). When you're at the beginning of your career, list education first.

When you're more experienced, move the education section after the experience section.
- Qualifications such as certification and/or license
- Professional experience, with job titles, dates of employment (month or year is fine), and a short, bulleted list of your duties and accomplishments. Be sure to include the things you accomplished or achieved in those roles.
- Military service (if any)
- Awards and honors (if any)
- Volunteer experience (if relevant)

Use a reverse chronological résumé when:

- Most of your experience has been in one field.
- Your work history demonstrates a clear career path.
- You work in a field that doesn't accept functional résumés.
- You want to include your résumé in an online jobs database or job search website.

Functional Résumé

A functional résumé is designed to highlight your skills and qualifications rather than your work history. Also called a skills résumé, the functional résumé demonstrates that you're a strong candidate for a job, deemphasizes periods of time when you weren't employed, and helps employers focus on specific skills needed for the job they're hiring for.

In a functional résumé, you break up your information into several categories that describe your skills. The categories should be in the order of most importance to the potential employer. Within each category, include a bulleted list of examples. These should also be in order of importance, rather than in chronological order.

Include a synopsis of your work experience. Even when you are leading with your skills, you still need to tell the interviewer what you've done in the past.

Sample: Reverse Chronological Résumé

Alex Tesla
13 Johnson Ave., Smallville TN 55555
Cell phone: 555-555-5555
Email: green.alex.37@email.net

RÉSUMÉ

Education
- AS in Electrical Technology, Smallville Community College, 2019
- Diploma, Smallville High School, 2015

Experience
Electrician's Assistant   2017–present
Fred's Electrical & Such, Sevierville, TN
- Assistant electrician with increasing responsibility
- Maintained tools in good working order and ensured right tools were available at the right time
- Basic electrical work: cutting, stripping, and bending electrical wire; testing for short circuits, loose connections, or missing insulation; making necessary repairs.

Roofer's Assistant   Summer 2014, 2015
Tesla Roofing & Repair
- Maintain and set up tools and scaffolding
- Clean roof and prep for roofing
- Attach roofing paper and roofing materials
- Clean work site

Other Achievements
- Red Cross First Aid certificate
- Second Prize, Smallville High School Science Fair for pico hydropower project

References available on request

The usual layout for a functional résumé is:

- Name and contact information at the top
- Summary of your skills and abilities
- Qualifications such as certification and licensing
- Awards and honors (if any)
- Relevant skill blocks in order of importance, such as technical skills, business skills, people skills
- Professional experience, with job titles and dates of employment (month or year is fine). Include short bulleted items about your duties and accomplishments if the jobs are different from each other in a significant way (otherwise you've already covered this in the skill blocks).
- Education, starting with most recent first (if you've been to college, you don't need to list your high school)
- Volunteer experience (if relevant)

Use a functional résumé when:

- You want to tailor your résumé to a particular job opening.
- You have less experience.
- You have highly specialized experience.
- You have gaps in your employment history.
- You have changed jobs frequently or after a short period of time.

For instance, if you spent the first twelve years of your career doing something else—like being an attorney or a kindergarten teacher or sailing around the world by yourself (or all three)—before deciding to enter an electrician program and apprenticeship, you might want to use a functional résumé to be sure your electrical skills are highlighted. You're not trying to hide the other things you've done; you're just making sure the employer can focus on what's relevant to the job in question.

Sample: Functional Résumé

Chris Traveller
100 Lane Blvd., Sunnyside VT, 05090
Cell: 555-555-5555
Email: solar.chris42@email.net

RÉSUMÉ

Summary
Licensed electrician experienced with installing residential and commercial solar panels. Team player with excellent customer service skills. Extensive knowledge of electrical and building codes.

Qualifications
- Journeyman electrician's license from the Electrical Licensing Board, VT Department of Public Safety, Division of Fire Safety, 2015
- Vermont Technical College Electrical Training and Apprenticeship Program, 2014

Skills Highlights
- Assembly and installation of solar photovoltaic panels
- Installation of complete electrical systems for homes and commercial buildings
- Troubleshooting and problem-solving for emergency repairs
- Blueprints and building plans
- Safety-conscious
- First Aid training
- Excellent math and report-writing
- Excellent time management
- Consistently positive customer reviews

Experience
- Solar PV Installer, Sunny Time Solar Panels   2017–2019
- Registered Electrical Apprenticeship with Sunnyside Solar   2014–2017
- Electrician's Assistant, Econolectrician LLC of Bennington   2010–2014

Education
- Continuing Education coursework in Solar, Solar Installation, Lightening Rod Speciality, and Electric Locksmith
- Vermont Technical College White River Junction Campus and distance learning, completed 2015
- BFA in Studio Art, Castle Rock College, Maine, 2005

Volunteer Work
- Guest speaker on electrical safety, Technical Students Association, Sunnyside Elementary–Middle School
- Habitat for Humanity

Combined Résumé

A combined résumé is the best of both worlds. It combines aspects of both the reverse chronological résumé and the functional résumé. A combined résumé is best if you have some experience, so that you have something to summarize. It summarizes your skills while still showing your impressive employment history.

Like a functional résumé, the combined résumé begins with a professional summary of your skills, abilities, and achievements that are specifically relevant to the job opening. Then your education and experience follow, in reverse chronological order.

The usual layout for a combined résumé is pretty simple:

- Name and contact information at the top
- Summary of your skills and abilities
- Qualifications such as certification and licensing
- Professional experience, with job titles and dates of employment (month or year is fine). Include short bulleted items about your duties and accomplishments if the jobs are different from each other in a significant way (otherwise you've already covered this in the skill blocks).
- Awards and honors (if any)
- Education, starting with your apprenticeship, then plumbing school, then high school
- Volunteer experience (if relevant)

Use a combined résumé when:

- You have a lot of experience and want to focus on your knowledge and accomplishments.
- You want to highlight your relevant experience.
- You're applying for a job that requires technical skills and expertise (like a clean energy technician).
- You want to move into a new field (such as clean energy).
- You want to demonstrate mastery in your field.

Sample: Combined Résumé

Aster McMaster
49 LaCienega Blvd., Tres Gatos CA 96165
Cell: 555-555-5555
Email: blow.wind.blow@email.net

– Résumé –

Summary
Highly qualified wind turbine technician with expertise in large, commercial wind farms.

Skills Highlights
- Electrical and hydraulic installation, troubleshooting, and maintenance
- Mechanical systems: generators, blades, braking, variable pitch systems, variable speed control systems, converter systems, and related components
- Computers and programmable logic control systems
- Drone inspection and thermal imaging techniques
- Project management
- Rescue, safety, first aid, and CPR training
- Team building and motivation

Experience
Los Gatos Wind Farm, Los Gatos, CA
Wind Turbine Technician III   2010–present
- Supervise staff of 11 wind turbine technicians
- Project manager for drone program
- Coordinate with management and administrative staff

Wind Turbine Technician II   2002–2010
- Maintained wind turbines

Solo Uno Wind Power
Wind Turbine Technician   2000–2002
Electrical Assistant   1998–2000

Education
- BS in Electrical Engineering, Los Gatos University   2019
- AA in electrical technology, Santa Boadicea City College   2000
- Continuing education hours

Community Service
- Los Gatos Water District Planning Board   2016–present
- Judge, Los Gatos High School Science Fair   2008, 2012, 2016

Writing Your Cover Letter

Your cover letter is an opportunity to tell your story. It's a short, personalized letter that you send with your résumé to introduce yourself to a potential employer. A well-written cover letter is a way to show a little of your personality, to highlight where and how your background makes you a good fit for the position you want, and to indicate your interest in working for that employer.

You should always try to send your letter and résumé together to the person who is responsible for making the hiring decision. If (and *only* if) you absolutely cannot find out who the decision maker is, send them to the human resources office.

Your letter should be in business letter format. (See image on page 96)

- Be sure your name and contact information are at the top of the letter, either centered or on the right.

Clean energy companies are hiring! You don't have to wait to see a job listing to let them know you're available. ©*Artur/iStock/Getty Images Plus*

- Address the reader by name—avoid generic greetings like "Dear Manager" or "Dear Director." Use Ms. or Mr. with the last name. (Do not use Miss or Mrs. unless you have been specifically instructed to do so.)
- Identify the specific position you are interested in and where you heard about it (some companies like to track how applicants heard about the position). Mention that your résumé is included or attached.
- If you heard about the opening from a specific person, mention him or her by name.
- Highlight your most relevant qualifications: skills that match the ones in the job description and/or skills that could transfer to those in the job description. Focus on your strengths and on what you could bring to the position. Think about this from the employer's point of view: What about your background will benefit them?
- Avoid negative language—phrase everything in a positive way. In particular, avoid complaining about a previous employer or customer.
- Your conclusion should include a confident call to action, such as requesting an interview. Don't ask directly for the job at this point, just an interview. Include your phone number here, as well as with your contact information at the top.
- Closing: Sincerely, (That's it. Don't use any other word.)
- Add a few lines of space for your signature, then type your name.
- Sign the letter by hand

Application Forms

Many companies offer an application form—either their own or a generic form provided by an online résumé or hiring service. As long as you keep your résumé up-to-date, you'll have all the information you need to quickly fill in these forms.

There are also online résumé sites like Monster.com and many others, where you can upload your résumé for employers to see. If you attend a two- or four-year college program, the Alumni Office may maintain a résumé bank like this, too.

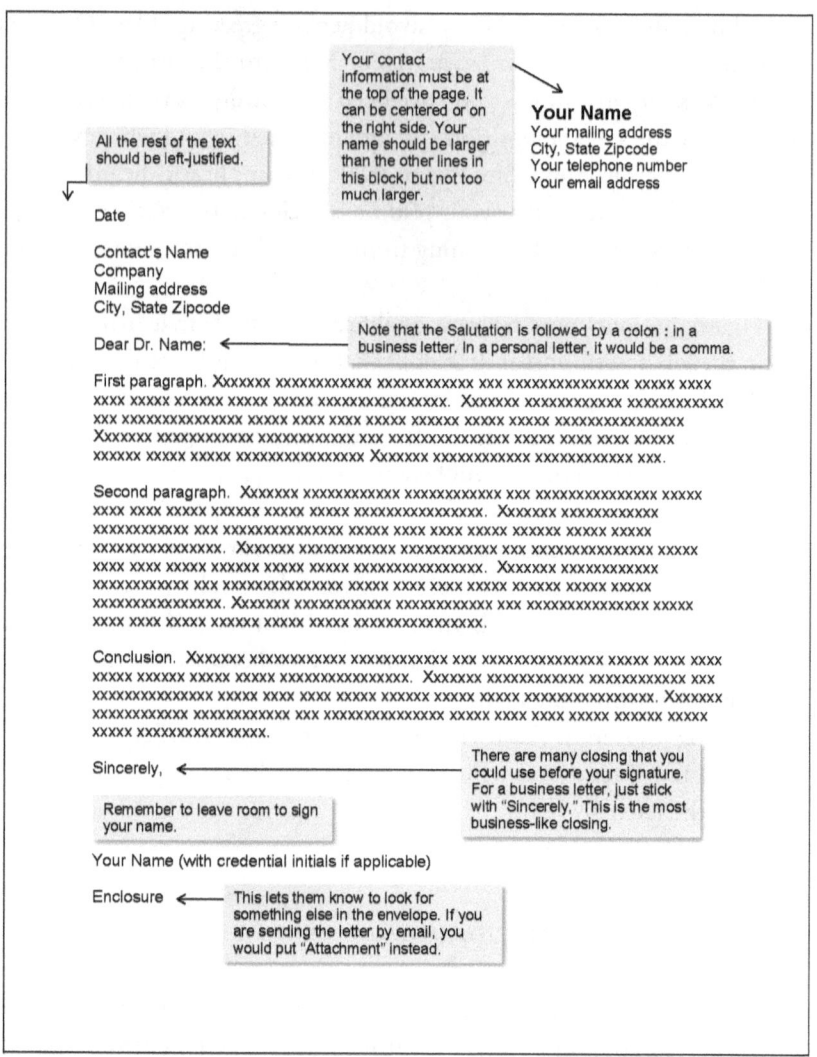

A Sample Cover Letter

ONLINE APPLICATION FORMS

Online application forms make things easier for applicants and employers. One advantage is that you can often copy and paste information from your résumé on your computer directly into the right box on the form.

Online forms can be a little unforgiving about what information they want and how they want you to provide it.

- Be sure to fill in all the boxes that are required—often marked with an asterisk (*).
- Fill in nonrequired boxes as best you can.
- If there is a place to upload your résumé, do that as well as filling in the form. If you can save your résumé as a PDF document, your formatting will be protected (the reader will see the document the same way that you do).
- Online forms are not perfect. For instance, if they ask for a letter, a résumé, and a list of references, but only give you a place to upload the résumé, try saving all the documents together (in that order) in a PDF file, then upload that file.
- Be sure you have filled in *everything* correctly before you hit SEND. Sometimes application forms will have a SAVE option so that you can come back to the form and make changes before you finally hit SEND.
- If you have any trouble with the form, call the company (usually the human resources or personnel office) and ask for help.

Paper Application Forms

Paper application forms are a lot less common than they once were, but you may still encounter them, especially at smaller companies. You'll need to copy information by hand from your résumé onto the paper form.

- Write neatly! Paper forms are read by people. Keep it in the boxes.
- Fill in all required information.
- Fill in as much nonrequired information as you have.
- See if you can attach your cover letter and your résumé.
- Paper application forms are usually filled in on site, at the potential employer's office, so if you have a problem or a question about the form, there may be someone you can ask. Be sure to ask!
- Be sure everything on the form is correct before you turn it in.
- If you make a mess of the form, with lots of changes and crossing out, ask for a new form and fill it in neatly.

Clean energy technician jobs are everywhere—you just need to look. ©*mstahlphoto/iStock/Getty Images Plus*

GETTING TO YES

There is a lot of work for clean energy technicians, and you will most likely find a job. But there's no guarantee that you will be offered the job you want most when you first start looking. Here are some tips that will improve your chances of getting to yes.

- Do your research—find out about the company that you want to apply to.
- Talk to people—especially people you know already or friends of friends who know something about that employer.
- Ask about what the potential employer is like to work for.
- Ask about what they value in their employees.
- Ask about benefits, and the general pros and cons of working there.
- If there is a specific job opening you're qualified for, apply for it!
- If there isn't a specific job opening, send a letter to the head of the company or the department you're interested in, mention your contacts, and ask if they would have a conversation with you about potential openings.
- Be flexible—you might find a good job in a different location than you wanted or doing something slightly different than you originally planned.
- Put your best self forward—everyone you meet is a potential contact for a job (or maybe just a new friend).
- If you get an interview, don't forget that all-important thank-you note! It's one of the most important things you can do to make a good impression. Send the note that day, as soon after the interview as possible.
- Don't put all your eggs in one basket—apply for numerous jobs at the same time.

The Interview

An interview is a business meeting where a prospective employer is checking you out. Don't forget that you are also checking them out. You are both there to see if it would be a good fit for you to work together. No matter how much you want the job, remember that you are not there to beg for charity—you are there to offer your services in your professional role.

DEALING WITH NO

A wise person once said, "If they didn't hire you, you probably would not have been happy working there anyway." Both employers and employees need to find the right fit. If they didn't think you were the right fit, you most likely wouldn't have thought so after a while, either. Here are some tips to get you through a no while you're waiting for the yes.

- Apply for lots of jobs at the same time, so no particular job will be too important to you.
- It doesn't feel great to be turned down for a job, but try not to take it personally.
- Don't burn your bridges! Don't retaliate with an angry letter or e-mail or troll the company all over social media. Another opportunity may come up there or with someone they know.
- Keep improving your résumé and your cover letter.
- Keep putting your best self forward—even if you're feeling discouraged, pick up your head and go through your day shining with confidence.
- Work your contacts—talk to other people you know. They may know an employer who would be a great match for you.
- Take advice—if someone (especially at or following an interview) tells you that you need to improve something, improve it. This may be an additional credential, or it may be something about your interpersonal skills or your spelling or your breath or whatever. If someone tells you something about yourself that you don't like to hear but suspect may be right, don't get mad—get better.
- Keep doing your research, so if one employer turns you down, you have three more to apply to that day.
- Keep telling yourself that employment is just around the corner. Then make it true!

As a clean energy technician, you are in high demand. But it's still important to make a good impression when you're applying for a job at any level. Be your best self, be confident, be polite.

WRITING YOUR RÉSUMÉ AND INTERVIEWING 101

Once you get the interview, you can show a prospective employer your value. ©*dusanpetkovic/iStock/ Getty Images Plus*

INTERVIEWING TIPS

- *Be on time.* Don't be late, *ever.* Try to arrive ten to fifteen minutes early so you have time to go into the restroom and check yourself in the mirror before you go into the interview. And don't be too early—that's just awkward.
- *Be polished.* See the section below on how to dress.
- *Bring your résumé.* Yes, they already have it. Bring extra copies just in case. It's helpful and shows that you're the kind of person who is prepared.
- *Smile.* Let them know that you will be a pleasant person to work with.
- *Shake hands well.* A firm handshake marks you as a person to be taken seriously. Shake hands as you come into the interview and again before you leave. (See the box "A Good Shake.")
- *Ask for a business card.* You may meet with just one person, with a committee, or with several people individually. At the end of the meeting,

ask for a business card from each person so that you have good contact information for your thank-you notes (see below).
- *Have good posture.* Sit up straight, make reasonable eye contact (not staring), and keep your shoulders back. Make it look normal, though—as though you always sit or stand that way. Good posture conveys energy and enthusiasm for the job, as well as showing you have the physical strength to do what clean energy technicians do.
- *Be prepared.* Learn about the company ahead of time so that you sound knowledgeable during the interview. Read the company's website and talk to people.
- *Be ready to answer questions.* At a job interview, you can expect to get asked some standard questions ("Where do you see yourself in five years?") and questions about the specific clean energy field (hydro, wind, solar, or geothermal) that show you know your stuff.
- *Don't be afraid to ask questions.* Some people don't like to ask questions in an interview because they think it makes them look ignorant. Actually, *not* asking questions makes them look uninterested. Have some questions prepared—both basic and more in-depth, because the basic ones might get answered before you have a chance to ask them.
- *Stay off your phone!* If you're looking at your phone during an interview, you'll look like you don't care. Nobody wants to hire someone who doesn't care before the job even starts!

A GOOD SHAKE

Here are the general rules of handshaking for a job interview:

- The person with more authority (the interviewer) should put out a hand first. If you accidentally go first, don't pull your hand back—that's rude.
- Face the other person, make eye contact, and smile.
- Be fairly still. You don't want to give the impression that you're trying to leave.
- Greet the person and say something pleasant, like, "Pleased to meet you." But don't gush.

A great handshake makes a great first impression. ©xavierarnau/E+/Getty Images

- Shake with your right hand unless it's injured. If the other person offers you their left hand, shake it with your right hand.
- A handshake is an up-and-down motion for usually about two or three seconds. Don't pump.
- A handshake should be firm—not limp and not crushing.
- If someone offers you a fist or an elbow to bump, go with it—they may be concerned about germs or they might have dropped a pipe on their hand a few minutes earlier.

You can't really learn how to shake hands by reading about it or watching a video. Those are good places to start, but you have to feel how it feels. The best way to learn to shake hands well is by asking someone who knows how to show you. How about your high school principal? She didn't get where she is today without knowing how to shake hands!

WHAT TO WEAR

As a clean energy technician, you'll wear hard-wearing clothes to work, like work pants, steel-toed boots, work shirts, coveralls, and safety equipment like a hard hat. So what should you wear to a job interview?

The answer is "business casual," which can mean different things for different places. You don't need to be fancy, but you should be presentable.

- If you're applying for an entry-level job with a smaller company, you could wear jeans and a nice button-down shirt or blouse. You can wear work boots instead of shoes, as long as they're clean.
- If you're applying for a job at a larger company, especially a larger business or a utility company, you should wear khakis instead of jeans, and a nice shirt or sweater. Men should add a tie, and women might wear some very simple jewelry. Wear nice but simple shoes.
- *Always* be clean and neat for an interview. Brush your teeth. Shower, wash your hair, have a fresh haircut or pull long hair back neatly. Avoid cologne and the smell of cigarettes. Don't chew gum.

> T-shirts may be appropriate for your workplace, especially worn under coveralls or a uniform. But message tees are never appropriate for work. Leave the ones with words, pictures, jokes, political or religious messages, cartoons, and the like at home. What's funny to you might be offensive to a customer, a coworker, or your boss—and being offensive is unprofessional.

What Potential Employers Look For

There are certain qualities that every clean energy technician should have. During a job interview, potential employers will be assessing you for these characteristics. Ask yourself these questions and if you think you need to get better at something, then get better!

COMMUNICATION AND SOCIAL SKILLS

- Will you be able to understand the customer's problems, needs, and values?
- Will you be able to work well with your boss and coworkers?
- Do you have active listening skills?
- Do you speak clearly?
- Do you write clearly?
- Do you show politeness, friendliness, and a good attitude?

GOOD WORK ETHIC

- Do you work hard at assigned tasks?
- Do you look for ways to help employers, coworkers, or customers beyond assigned tasks?
- Do you look for ways to improve your performance?
- Are you on time?
- Do you show initiative and work to solve problems?

ADAPTABILITY

- Are you flexible about new situations, new rules and regulations, and new or different environments?
- Are you willing and eager to learn the latest developments, processes, procedures, and code updates?
- Can you get along with all kinds of people?

ENTHUSIASM FOR YOUR FIELD

- Are you proud of the work you do as a clean energy technician?
- Do you like solving problems?
- Do you like helping people?
- Do you have a desire to continue to build your skills and learn new things?

Following Up

After any kind of job interview, it is *extremely* important to follow up. This is what shows the interviewer that you are genuinely interested in the job and in working with the company. Write your thank-you note immediately after the interview. Be sure to mention your interest in the job and one or two things from the interview that interested you most. If you met separately with several people, send each one of them a separate note.

HANDWRITTEN LETTER

Traditionally, a handwritten letter has been the gold standard for thank-you notes. (If your handwriting is truly terrible, then type the letter and sign it by hand.) These must be mailed the same day as your interview, so be sure you already have stamps and envelopes before you even go to the interview.

```
Your name
Your mailing address                                       Stamp
City, state, zip


                                   Name
                                   Company
                                   Street Address
                                   City, state, zip
```

How to address an envelope

E-MAIL THANK-YOU NOTE

While an e-mail is less personal than a hand-signed letter on paper, it's a lot faster and is considered an acceptable way to communicate. An e-mail should have all the same content as a handwritten letter, including (and especially) communicating your enthusiasm for the job.

Just like a handwritten note, start with "Dear Ms. Name" (replacing "Name" with whatever the person's name is) and sign it "Sincerely," two line breaks, "Your Name." Since you won't be writing your signature, one blank line is enough.

> Many employers require that employees be able to pass a drug test and a background check. Think about it—nobody wants to find out too late that their employee was impaired while working, endangering themselves and everyone around them. Employers need to be sure that their employees are responsible people, especially since there are a lot of safety issues as well as liability issues involved in clean energy and electrical work. So be sure that you can meet those requirements!

On the Job

Now that you've got the job, it's important to keep it! It's not that hard. Just remember these simple tips:

- *Safety first.* Clean energy technicians deal with hazards all day long. Stay safe and watch out for the safety of those around you.
- *Do your best.* Your biggest asset is high-quality work.
- *Be reliable.* Your coworkers, clients, and supervisors will respect and appreciate you most when they know they can rely on you.
- *Be on time.* Show up on time for work or even a few minutes early.
- *Be prepared.* Walk in the door ready to work.
- *Keep good records.* This is important for safety as well as efficiency.
- *Be polite.* Treat everyone you meet with the same respect you want to receive.
- *Stay calm.* You do your best work when you're calm, especially if there's a problem to solve or an emergency.
- *Have integrity.* Be honest and respect other people's person and property.

Safety is the most important part of a clean energy technician's job. ©CharlieChesvick/E+/Getty Images

It looks like you're ready to get started on your journey to be a clean energy technician! ©michaeljung/ iStock/Getty Images Plus

Summary

There is a strong market out there for all the types of clean energy technicians we've talked about in this book, and there will continue to be. As long as you do a good job and respect the people you work with and for, you should have a long and lucrative career. You'll be figuring out mysteries, solving problems, working with your brain and your hands, and most important—making our world safer and better for everyone.

Good luck!

WAYNE KILCOLLINS—WIND TURBINE TECHNICIAN

Wayne Kilcollins is an instructor in wind power technology at Northern Maine Community College (NMCC) in Presque Isle, Maine. He holds a BS in mechanical engineering and an MBA from Century University in Albuquerque, New Mexico.

Wayne Kilcollins has the best view in Presque Isle from the top of a three-hundred-foot wind turbine.
Photo by Brian McDougal

He was a Technician Level 3 and then environmental health and safety coordinator with General Electric (GE) Power and Water Division after a previous career as an engineering project manager for a manufacturing company.

How did you decide to become a wind turbine technician?

I was laid off from an engineering project manager position due to the relocation of manufacturing to offshore facilities. I had spent more than eighteen years coordinating automated production equipment development and construction for an electronic components manufacturer. When the company relocated their production facility to the Dominican Republic, I spent the last several years relocating the equipment along with training operators and technicians to work with the equipment. Once they were up to speed and the equipment met quality and production requirements, I was laid off. I was given the opportunity to move along with the company but decided to stay in northern Maine. So I began the search for another interesting career. I was made aware of a GE wind technician opening in a nearby town. I was interested in production operations and troubleshooting equipment from my earlier career activities, so I decided to look further into the opportunity. The opening was for a GE operations and maintenance contract at a wind farm. I applied for the position and was invited to interview a couple of weeks later. The interview process included a face-to-face with the regional manager, multiple technical competency exams, and a climb test in a 1.5-megawatt wind turbine. The exams were challenging, but my interest was in the eighty-meter climb up the wind turbine. Wow! What a view from the top. I could see most of northern Maine from the top of the nacelle.

What is a typical day on the job like for a wind turbine technician?

For wind turbine technicians, no day is the same. When you get to the site at six or seven in the morning, you're given work orders. You could be doing troubleshooting on equipment with electronics issues or you could be doing some regular maintenance. For maintenance, you have to shut the equipment down, but it all depends on the weather. If the winds are good, they'll say keep the equipment running. It's a generator, feeding electricity to the grid, so if it's running, you don't shut it down. But if the winds aren't blowing, then you do.

What's the best part of the job?

I like a challenge. You take a look and see what the issues might be, go through the information that the turbine provides. SCADA is the supervisor control and data acquisition system. It gives you diagnostics, just like a doctor would do if you went to the hospital. You can look at the data and get a lot of the diagnostics for what you need to do for troubleshooting before getting into the machinery. Also, when you step on top of one of those turbines, it's the best view in the world!

What's the most challenging part of the job?

Some people would say it's the climb! After you do it a few times, it's not so bad. For a lot of students when they first come here, there's a lot of apprehension about going three hundred feet in the air up a ladder.

What's the most surprising thing about the job?

One of the things I thought was really neat when I was working for GE up in northern Maine was we had a lot of traveling technicians who'd come in to help the local team. We'd have people from Texas or New Jersey or New York. You'd share experiences with what you've seen on the site and what they've seen in their travels. There's a lot of camaraderie with the other technicians. I went to Sweetwater, Texas, for training and out to Schenectady, New York, for troubleshooting and operations training. Even on the training, you meet a lot of technicians from a lot of locations and you hear what their experiences are. You share experiences much more than in a typical manufacturing facility, for instance.

How did (or didn't) your education prepare you for the job?

My education and background in engineering allowed me to understand how the systems work together. Whether it was electronics, hydraulics, or mechanical systems, I had an understanding of it, so that made it easier to do troubleshooting. It also made it easier working with my students. Engineering and manufacturing are very similar, but it's just that your office isn't three hundred feet in the air.

Is the job what you expected?

It is. The type of equipment was similar, so I expected the kinds of electronic and mechanical troubleshooting you do as a wind turbine technician. I had experience in manufacturing as well as research and development.

How did you decide to begin teaching?

I had been on some committees at Northern Maine Community College, so when I finished my master's degree, the department head asked me to interview for the wind power program that was just starting up. Now I teach full-time. I teach the related courses for the wind turbine technician program at the college. I have a 200-kilowatt turbine on a four-foot tower in my classroom. My boss said whatever I could fit through the back door, I could have! It's great to have a functioning unit in the lab that the students can work on and learn on without having to climb a tower to get up to it. So they can all gather around and see how things are done and to share the experience before they go and do it.

What's the biggest challenge your students face?

Trying to grasp both mechanical and electrical concepts. Some students come in with a real passion for electronics systems and others with a passion for mechanical systems. They need to have a decent skill set in both to go out to work in the field as efficient technicians and troubleshooters. The ones who are mechanical have to step out of their comfort zone to learn electrical and vice versa.

How are you using drones in your program at NMCC?

Drones are a new addition in the last couple of years. Manufacturers have specialty teams that go out to do inspections on the turbines. The drones let them fly up the outside of the unit and inspect the blades and the tower to see if there's any damage that needs to be repaired. Then they can pass that information along to someone to do the repair. With the drones, they don't have to put a technician at risk to rappel down the tower to find the damage, so there's only one climb instead of two. It increases the safety and reduces the risk. I've got three drones, which give the students the hands-on skills. I go through the materials they need for FAA certification. They take the online course with the FAA and then take the certification exam.

Where do you see the career of wind turbine technician going from here?

The graduates I have going out in the field are working in development (developing the site—where's a good place to put the wind farm), manufacturing, and maintenance. As it goes forward, there are so many more systems being constructed in the United States, Canada, and other locations. With some of these manufacturers, there are a couple hundred slots available. We can't get students through the program fast enough. If you look at AWEA [American Wind Energy Association] and some of these other sites, you'll see that over the next seven or eight years, they're looking for hundreds or thousands of technicians. If they want a job, they can get a job. If they want to travel, they'll have the opportunity. The pay for travel is more than for on-site, plus they pay a per diem and traveling expenses, and provide a vehicle. The field is really exploding. I've been with the college eleven years, and I have graduates from Maine to Hawaii. Some of them have been promoted to site lead or site manager, which is great in such a short amount of time. In other industries you have to be there quite a while before you get into those lead positions.

What is your advice for a young person considering this career?

Study algebra. Be comfortable with math. I know a lot of students who come through my program or electrical construction or computer science (the other programs in my department) who start off thinking math isn't a big deal, then discover they have to spend time figuring stuff out. It gives them the logic to go through the step-by-step processes you need for troubleshooting. If they can catch on to the manual skills of doing the job, it helps to understand the math behind it. It's not rocket science—it's just algebra. The ones who think they don't need it in high school are short-changing themselves.

NMCC has a great program in wind power technology! I've had students that come in with associate's degrees and bachelor's degrees in lots of different fields. They're looking for something where they can see what they've done at the end of the day. I've had students as far away as Maryland and Tennessee. My lab assistant is from the UK. When he's done, he wants to get the experience here in the United States and then go back to Europe. I had a student from Finland whose dad was working in Massachusetts who came up here to take the program. We have a great relationship with a lot of manufacturers and developers, so if someone is really interested in a career in wind, I don't know a lot of other places that have a turbine sitting in the classroom.

My philosophy in teaching is see it, do it, teach it. So I have them see how a process is done, then do it themselves, then teach one of their classmates. Understanding how to explain it to somebody else really brings the concept home. I think in any technical job, that's a really good way for them to grasp it and understand the whole process. It helps to explain the details to somebody else. When the students share their experiences, it really helps with the whole process.

Notes

Introduction

1. Gaylord Nelson, "Environment, Population, Sustainable Development: Where Do We Go from Here?" remarks on the 25th Anniversary of Earth Day, April 22, 1995. Available at http://www.nelsonearthday.net/nelson/quotes.php.
2. Avery Phillips, "Career Pathways and Possibilities in Renewable Energy," *Renewable Energy World*, February 28, 2018, https://www.renewableenergyworld.com/2018/02/28/career-pathways-and-possibilities-in-renewable-energy/.
3. US Department of Energy Office of Energy Efficiency & Renewable Energy, "5 of the Fastest Growing Jobs in Clean Energy," April 24, 2017, https://www.energy.gov/eere/articles/5-fastest-growing-jobs-clean-energy.
4. Frank Jossi, "Wind-Solar Pairing Cuts Equipment Costs While Ramping Up Output," on *Renewable Energy World*, March 11, 2019, https://www.renewableenergyworld.com/2019/03/11/windsolar-pairing-cuts-equipment-costs-while-ramping-up-output/.

Chapter 1

1. Lora Shinn, "Renewable Energy: The Clean Facts," National Resource Defence Council, June 15, 2018, https://www.nrdc.org/stories/renewable-energy-clean-facts.
2. Mari Hernandez, "2020 Clean Energy States Honor Roll," Interstate Renewable Energy Council, January 22, 2020, https://irecusa.org/2020/01/2020-clean-energy-states-honor-roll/.
3. Trade-Schools.net, "Renewable Energy Jobs: Careers That Make a Difference in the World," https://www.trade-schools.net/articles/renewable-energy-jobs.asp#solar.
4. Environmental and Energy Study Institute (EESI), "Fact Sheet—Jobs in Renewable Energy, Energy Efficiency, and Resilience (2019)," July 23, 2019, https://www.eesi.org/papers/view/fact-sheet-jobs-in-renewable-energy-energy-efficiency-and-resilience-2019.
5. American Wind Energy Association, "Wind 101: Basics of Wind Energy," https://www.awea.org/wind-101/basics-of-wind-energy.

6. Christine Real de Azua, "U.S. Wind Resource Even Larger than Previously Estimated: Government Assessment," American Wind Energy Association [press release], February 18, 2010, https://web.archive.org/web/20100224133904/http://www.awea.org/newsroom/releases/02-18-10_US_Wind_Resource_Larger.html.

7. US Department of Energy, Office of Energy Efficiency & Renewable Energy, "Everything You Need to Know about Wind Turbine Technicians," October 23, 2017, https://www.energy.gov/eere/articles/everything-you-need-know-about-wind-turbine-technicians.

8. OwlGuru.com, "How Much Do Wind Turbine Technicians Make in 2018?" https://www.owlguru.com/career/wind-turbine-service-technicians/salary/.

9. US Bureau of Labor Statistics *Occupational Outlook Handbook*, "Wind Turbine Technicians: Job Outlook," https://www.bls.gov/ooh/installation-maintenance-and-repair/wind-turbine-technicians.htm#tab-6.

10. Ibid.

11. Union of Concerned Scientists, "Renewable Energy: Unlimited Resources. Little or No Pollution. Solar Power," https://www.ucsusa.org/energy/renewable-energy.

12. Jacob Marsh, "What Is Solar Energy?" EnergySage, January 10, 2010, https://news.energysage.com/what-is-solar-energy/.

13. US Bureau of Labor Statistics *Occupational Outlook Handbook*, "Solar Photovoltaic Installers: Pay," https://www.bls.gov/ooh/construction-and-extraction/solar-photovoltaic-installers.htm#tab-5.

14. Solar Foundation, "Solar Ready Vets Network," https://www.thesolarfoundation.org/solar-ready-vets/.

15. Career Explorer, "Hydroelectric Plant Technician Salary," https://www.careerexplorer.com/careers/hydroelectric-plant-technician/salary/.

16. Anh Speer, "Hydroelectric Plant Technician Jobs and Green Career Profile," Renewal Energy Jobs, January 30, 2014, http://www.isustainableearth.com/green-jobs/renewable-energy-job-profiles/hydroelectric-plant-technician-jobs.

17. Tennessee Valley Authority, "Careers: Hydro Technician Training Program," https://www.tva.com/Careers/Hydro-Technician-Training-Program.

18. Career Explorer, "The Job Market for Hydroelectric Plant Technicians in the United States," https://www.careerexplorer.com/careers/hydroelectric-plant-technician/job-market/#how-employable.

19. US Energy Information Administration, "Geothermal Explained: Where Geothermal Energy Is Found," https://www.eia.gov/energyexplained/geothermal/where-geothermal-energy-is-found.php.

20. US Bureau of Labor Statistics *Occupational Outlook Handbook*, "Heating, Air Conditioning, and Refrigeration Mechanics and Installers: Job Outlook," https://www.

bls.gov/ooh/installation-maintenance-and-repair/heating-air-conditioning-and-refrigeration-mechanics-and-installers.htm#tab-6.

Chapter 2

1. Jared Diamond, *Collapse: How Societies Choose to Fail or Survive* (New York: Penguin, 2011), 522.

2. Yogi Berra, *When You Come to a Fork in the Road, Take It!* (New York: Hyperion, 2001), 55.

3. Steve Maraboli, *Life, the Truth, and Being Free* (Port Washington, NY: Better Today, 2009), 26.

4. PayScale, "Average Wind Turbine Technician Hourly Pay," https://www.payscale.com/research/US/Job=Wind_Turbine_Technician/Hourly_Rate.

5. US Bureau of Labor Statistics *Occupational Outlook Handbook*, "Wind Turbine Technicians," https://www.bls.gov/ooh/installation-maintenance-and-repair/wind-turbine-technicians.htm.

6. PayScale, "Average Solar Energy System Installer Hourly Pay," https://www.payscale.com/research/US/Job=Solar_Energy_System_Installer/Hourly_Rate.

7. US Bureau of Labor Statistics *Occupational Outlook Handbook*, "Solar Photovoltaic Installers," https://www.bls.gov/ooh/construction-and-extraction/solar-photovoltaic-installers.htm.

8. Environmental and Energy Study Institute (EESI), "Fact Sheet—Jobs in Renewable Energy, Energy Efficiency, and Resilience (2019)," July 23, 2019, https://www.eesi.org/papers/view/fact-sheet-jobs-in-renewable-energy-energy-efficiency-and-resilience-2019.

9. Union of Concerned Scientists, "Achieving 100 Percent Clean Electricity," December 4, 2018, https://www.ucsusa.org/resources/achieving-100-percent-clean-electricity.

10. North American Electric Reliability Corporation (NERC), "System Operator Certification," https://www.nerc.com/pa/Train/SysOpCert/Pages/default.aspx.

Chapter 3

1. B. B. King, quoted outside the Main Library in uptown Charlotte, North Carolina, *Charlotte Observer*, October 5, 1997, 2D.

2. US Bureau of Labor Statistics *Occupational Outlook Handbook*, "How to Become a Wind Turbine Technician," https://www.bls.gov/ooh/installation-maintenance-and-repair/wind-turbine-technicians.htm#tab-4.

3. North American Board of Certified Energy Practitioners, "NABCEP Board Certifications," https://www.nabcep.org/certifications/nabcep-board-certifications/.

4. North American Board of Certified Energy Practitioners, "Associate Program," https://www.nabcep.org/certifications/associate-program/.

5. North American Board of Certified Energy Practitioners, "NABCEP Course Catalog," https://coursecatalog.nabcep.org/.

6. College Board, "Average Published Undergraduate Charges by Sector and by Carnegie Classification, 2018–19," Trends in Higher Education, https://trends.collegeboard.org/college-pricing/figures-tables/average-published-undergraduate-charges-sector-2018-19.

Chapter Four

1. Jeff Butterfield, *Written Communication*, 3rd ed. (Boston: Cengage, 2017), p. 80.

Glossary

Associate's degree: An academic degree granted after a two-year course of study, usually from a community/junior college, technical college, or trade school.

Bachelor's degree: An academic degree awarded to a person by a college or university after completion of an undergraduate degree program (usually four years); also called a baccalaureate degree.

BTU: Abbreviation for British thermal unit, which was originally defined as the amount of heat required to raise the temperature of one pound of water by 1 degree Fahrenheit. Because of variability in the way water changes temperature, there are now several different ways to define a BTU.

Clean energy: Electricity created using nonpolluting sources, specifically sources that do not emit carbon into the atmosphere. Sources of clean energy include solar, wind, moving water, plants, algae, and geothermal heat.

Clean energy technician: A worker with the training and expertise to install, maintain, and repair clean or renewable energy equipment and technology.

Continuing education: Coursework beyond a degree or certificate program. Continuing education is usually required to maintain certification or for certain types of employment to ensure that a worker knows the most recent and relevant information relating to their job.

Cover letter: Business letter that goes with another document (such as a résumé) to explain the contents and provide more context.

GED: Abbreviation for general equivalency diploma. A GED is earned by taking a series of academic tests that certify the (American or Canadian) test taker has achieved high school level academic skills. Considered to be an equal credential to a high school diploma.

Geothermal heating systems: Heat drawn from four to six feet below ground by means of coiled pipes and transferred directly into a building, such as a home, to provide heat directly.

Geothermal loop: Underground pipes made of high-density polyethylene that are used to move heat to and from the earth in geothermal heating systems. Open loop systems make use of natural groundwater, drawing it from underground and returning it there. Closed loop systems have water sealed within the circuit with a small amount of environmentally friendly antifreeze.

Geothermal power: Power generated using the natural heat of the planet.

Geothermal power plant: A power plant where electricity is generated by accessing hot water as liquid or steam from deep underground and using it to move a turbine to create electricity.

Geothermal technician: A worker with the training and expertise to install, maintain, and repair geothermal heating systems; a worker with the training and expertise to maintain and repair geothermal power plant technology.

Gigawatt: One billion watts.

Green energy: Energy from clean and renewable resources, including sunlight, wind, moving water, plants, algae, and geothermal heat.

Grid (or "the grid"): Short for the electrical grid (i.e., the electrical power system network) in a particular area. The grid connects facilities that generate electricity to those that use it. Some clean energy technologies allow home-generated power to be fed back into the grid, to be used when needed (*see* **net metering**).

HVAC: Abbreviation for heating, ventilation, and air conditioning; used to describe the technology that affects the indoor temperature and air quality of buildings and vehicles.

Hydroelectric plant: A power plant that generates electricity by using moving water to turn turbines.

Hydroelectric power: Electricity generated by using moving water to turn a turbine.

Hydro technician: A worker with the training and expertise to maintain and repair hydropower generators and other equipment in the hydropower plant.

Kilowatt: One thousand watts.

Kinetic: Relating to or resulting from motion.

Megawatt: One million watts.

Microhydropower: A small hydropower system that can generate up to 100 kilowatts of electricity.

Mini split: A ductless heat pump that blows heated or cooled air directly into a room; generally used for smaller spaces rather than entire homes.

Nacelle: A covering or housing that contains all the electricity-generating components of a wind turbine, including the generator, gearbox, drive train, and brake assembly.

Net dependable capacity: The amount of power a dam can produce on an average day, minus the electricity used by the dam itself.

Net metering: A system in which surplus renewable energy generated at a home or other location is transferred to a utility's electrical grid, which offsets the cost of electricity when that customer draws power from the grid.

Photovoltaic: Direct conversion (at the atomic level) of sunlight into electricity by capturing free electrons and the resulting electric current.

Photovoltaic cell: *See* **solar cell**.

Pico hydroelectric plant: Hydroelectric generators that generate no more than 5 kilowatts of electricity; sometimes used in small, remote communities that need only a small amount of electricity.

Power plant: Any facility that generates utility-level quantities of electricity, usually measured in megawatts or gigawatts.

Rappel: Descending a near-vertical surface by using a doubled rope coiled around the body and fixed at a higher point.

Renewable energy: Electricity generated from clean energy sources that naturally replenish themselves and therefore cannot be used up. Examples include solar, wind, water, and geothermal power.

Résumé: A brief written list of an individual's professional and educational history.

Solar cell: An electrical device that converts the energy of light directly into electricity by the photovoltaic effect that can produce a maximum open-circuit voltage of approximately 0.5 to 0.6 volts. Groups of solar cells can be joined to form solar panels. Also called a **photovoltaic cell**.

Solar installer/technician: A worker with the training and expertise to install, maintain, and repair solar panels and solar panel arrays.

Solar panel: A packaged, connected assembly of photovoltaic solar cells arranged into an array that generates electricity for commercial and residential applications.

Solar power: Electricity generated from sunlight via photovoltaic solar cells, concentrated solar power, or a combination.

Solar thermal capture: Capturing heat directly from solar radiation to use in a variety of ways. Low-, mid-, and high-temperature solar thermal energy systems are used for different purposes. Less practical than photovoltaic (PV) systems for small-scale electricity generation.

Turbine: A machine that transforms rotational energy from a fluid (such as water, steam, gas, or air) that is picked up by a rotor system into usable work or energy.

Watt: A unit used to measure electricity; a watt refers to the amount of electricity being used at a specific moment.

Watt-hour: A unit used to measure how fast electricity is used over the period of one hour.

Wind farm: An area of land where a group of wind turbines is situated to generate electricity, usually at a utility level. There is no specific number of turbines required for a wind farm.

Windmill: A machine that uses the power of the wind to turn blades to create mechanical energy; windmills have been used for centuries to grind grain and pump water. Windmills do not generate electricity.

Wind power: Electricity produced by using the power of the wind to turn a turbine.

Windtech: Abbreviation for **wind turbine technician**.

Wind turbine: A device that converts the kinetic energy of the wind into electrical energy by turning two or three blades around a rotor connected to a shaft that spins a generator; also called a wind energy converter.

Wind turbine technician: A worker with the training and expertise to maintain and repair wind turbines.

Zero-emissions: Refers to an energy source, engine, motor, or process that does not release any pollutants into the environment.

Resources

This section includes useful resources relating to clean energy technician careers. While this is not a complete list of all the information out there, these resources will help you to get started finding out more about the clean energy field you're interested in.

Resources for Clean Energy Technicians

Clean Energy Jobs List
https://www.cleanenergyjobslist.com
Information on jobs in clean energy careers. You can explore different companies, browse available jobs, or post your professional profile so companies can find you. The site also includes blog posts with new and relevant information on different aspects of the clean energy sector.

Electronics Technicians Association, International (ETA)
https://www.eta-i.org/renewable_energy.html
Offers renewable energy certification in:

- Photovoltaic Installer—Level 1 (PVI 1)
- Photovoltaics Installer/Designer (PV2)
- Small Wind Installer (SWI)
- Electric Vehicle Technician (EVT)

National Resources Defence Council (NRDC)
https://www.nrdc.org
A nonprofit membership organization that "works to safeguard the earth—its people, its plants and animals, and the natural systems on which all life depends." Its work covers climate change, communities, energy, food, health, oceans, water, and the wild.

The Solar Foundation Solar Ready Vets Network
https://www.thesolarfoundation.org/solar-ready-vets/
The Solar Ready Vets Network is funded by the U.S. Department of Energy Solar Energy Technologies Office and follows a two-track approach. The two major initiatives are the Solar Ready Vets Fellowship, and the Solar Opportunities and Readiness (SOAR) Initiative. From their website:

> Veterans of the US Armed Forces are outstanding candidates for careers in the solar industry. Military service provides the leadership abilities and technical skills that solar companies value highly. While some veterans begin with entry-level jobs and move up the ranks, others transition directly to advanced leadership roles within the rapidly growing solar workforce.
>
> Through the Solar Ready Vets Network, the Solar Foundation is connecting transitioning military service members and veterans with solar industry career training and professional development opportunities. In partnership with the U.S. Chamber of Commerce Foundation's "Hiring Our Heroes" program, the North American Board of Certified Energy Practitioners (NABCEP), and the Solar Energy Industries Association, they are expanding and strengthening a nationwide pipeline of military talent into a range of technical and leadership roles in the solar industry.

Student Energy
https://www.studentenergy.org
A global charity building the next generation of energy leaders to accelerate the world's transition to a sustainable energy future. Student Energy engages youth in unique programs that empower them to become change agents and work with actors within the energy system to create space for youth to have an impact.

US Bureau of Labor Statistics Occupational Outlook Handbook
https://www.bls.gov/ooh/
A searchable online database with extensive information on hundreds of jobs and careers, including different kinds of jobs in given fields, salary information, education and training requirements, projected demand, and much more.

US Department of Energy Office of Energy Efficiency & Renewable Energy
https://www.energy.gov/eere/
Provides graphics that map career opportunities in different sectors of clean energy. Each map shows different career areas for entry-level, mid-level, and

advanced skill levels; both broad and detailed views of jobs; routes to advancement; and frequently asked questions.

- Bioenergy Career Map—https://www.energy.gov/eere/bioenergy/bioenergy-career-map
- Hydrogen and Fuel Cells Career Map—https://www.energy.gov/eere/fuelcells/hydrogen-and-fuel-cells-career-map
- Solar Career Map—https://irecsolarcareermap.org/
- Wind Career Map—https://www.energy.gov/eere/wind/wind-career-map

Professional Organizations

Membership in a professional organization is a way to connect with other people in your field, learn the latest information through conferences and publications, and demonstrate your commitment to the field. Certification from a professional organization indicates you have the skills and training necessary to be a clean energy technician.

WIND TURBINE TECHNICIAN

American Wind Energy Association (AWEA)
https://www.awea.org/
The national trade association for the US wind industry. AWEA promotes wind energy as a clean source of electricity for American consumers. Members receive extensive benefits and informative resources.

Global Wind Energy Council
https://gwec.net/
The international trade association for the wind power industry. Many wind farms are owned by international companies. Sponsors programs such as Women in Wind Global Leadership Programme (https://gwec.net/women-in-wind/apply/).

SOLAR PHOTOVOLTAIC INSTALLER

International Solar Energy Society (ISES)
https://www.ises.org/
The largest international solar organization, with individual and corporate members in more than 110 countries. ISES works with the United Nations on important events like the UNFCCC Climate Change Conferences and the UN Commission on Sustainable Development meetings and provides objective scientific advice to governments and the public at a global level.

North American Board of Certified Energy Practitioners (NABCEP)
https://www.nabcep.org/
A nonprofit organization offering professional certification and accreditation for individuals and companies for photovoltaic system installers, solar heat installers, technical sales, and other renewable energy professionals.

Solar Energy International (SEI)
https://www.solarenergy.org/
Offers solar professionals certificate programs in residential and commercial photovoltaic systems, battery-based photovoltaic systems, solar business and technical sales, international and developing world applications, renewable energy applications, and solar professionals trainer.

HYDROPOWER PLANT TECHNICIAN

International Hydropower Association (IHA)
https://www.hydropower.org/
A nonprofit international membership organization committed to advancing sustainable hydropower. Formed under the auspices of UNESCO to promote and disseminate good hydropower practice, IHA champions continuous improvement and sustainable practices across the sector. The organization offers both individual and corporate membership.

National Hydropower Association (NHA)
https://www.hydro.org/
A nonprofit association dedicated to promoting the growth of clean, renewable hydropower and marine energy. NHA works to influence state and federal energy policy, provides industry information and access, and programs such as:

- NHA Job Board—https://www.hydro.org/job-board/
- Waterpower Resource Library—https://www.hydro.org/resources/
- Women in Hydropower Mentorship Program—https://www.hydro.org/resources/women-in-hydro-mentorship/

GEOTHERMAL TECHNICIAN

Geothermal Resources Council (GRC)
https://geothermal.org/
A nonprofit educational association that serves as a primary professional educational association for the international geothermal community. GRC provides continuing professional development for its members through its outreach, information transfer, and education services. The former Geothermal Energy Association (GEA) is now part of the GRC.

International Ground Source Heat Pump Association (IGSHPA)
https://igshpa.org/
A nonprofit member organization to advance ground source heat pump (GSHP) technology on local, state, national, and international levels. Headquartered on the campus of Oklahoma State University in Stillwater, Oklahoma, IGSHPA offers membership benefits and training.

North American Energy Reliability Corporation (NERC)
https://www.nerc.com/
Certifies geothermal technicians whose jobs could affect the power grid.

Learn More About Clean Energy

Clean Energy, US Department of Energy
https://www.energy.gov/science-innovation/clean-energy

Green Power Partnership, US Environmental Protection Agency
https://www.epa.gov/greenpower

Bibliography

Alliant Energy Corp. "Geothermal Energy." Alliant Energy Kids. https://www.alliantenergykids.com/RenewableEnergy/GeothermalEnergy.

Alvarez, Greg. "The Truth about Wind Power." *Into the Wind: The AWEA Blog*, April 10, 2019. https://www.aweablog.org/the-truth-about-wind-power/.

American Wind Energy Association. "Wind 101: Basics of Wind Energy." https://www.awea.org/wind-101/basics-of-wind-energy.

Berra, Yogi. *When You Come to a Fork in the Road, Take It!* New York: Hyperion, 2001.

Bruce, Leilani. "Top Careers in Renewable Energy." Everglades University, December 4, 2018. https://www.evergladesuniversity.edu/5-renewable-energy-careers-in-high-demand/.

Calpine Corporation. "About Geothermal Energy: Welcome to The Geysers." http://geysers.com/geothermal.

Career Explorer. "How to Become a Geothermal Technician." https://www.alliantenergykids.com/RenewableEnergy/GeothermalEnergy.

———. "Hydroelectric Plant Technician Salary." https://www.careerexplorer.com/careers/hydroelectric-plant-technician/salary/.

———. "The Job Market for Hydroelectric Plant Technicians in the United States." https://www.careerexplorer.com/careers/hydroelectric-plant-technician/job-market/#how-employable.

———. "What Does a Geothermal Technician Do?" https://www.careerexplorer.com/careers/geothermal-technician/.

Career One Stop. "Green Careers: Renewable Energy Generation." https://www.careeronestop.org/GreenCareers/ExploreGreenCareers/renewable-energy.aspx.

Career Planner. "Geothermal Technician." https://job-descriptions.careerplanner.com/Geothermal-Technician.cfm.

Diamond, Jared. *Collapse: How Societies Choose to Fail or Survive.* New York: Penguin, 2011.

Environmental and Energy Study Institute. "Fact Sheet—Jobs in Renewable Energy, Energy Efficiency, and Resilience (2019)," July 23, 2019. https://www.eesi.org/papers/view/fact-sheet-jobs-in-renewable-energy-energy-efficiency-and-resilience-2019.

Hernandez, Mari. "2020 Clean Energy States Honor Roll." Interstate Renewable Energy Council, January 22, 2020. https://irecusa.org/2020/01/2020-clean-energy-states-honor-roll/.

Illinois WorkNet Center. "Hydroelectric Plant Technicians." https://apps.il-work-net.com/cis/clusters/OccupationDetails/100576?parentId=111300&section=conditions&sectionTitle=Working%20Conditions.

Jossi, Frank. "Wind-Solar Pairing Cuts Equipment Costs While Ramping Up Output." *Renewable Energy World*, March 11, 2019. https://www.renewableenergyworld.com/2019/03/11/windsolar-pairing-cuts-equipment-costs-while-ramping-up-output/.

King, B. B. Quoted in *Charlotte Observer*, October 5, 1997, 2D.

Knier, Gil. "How Do Photovoltaics Work?" NASA Science: Share the Science. https://science.nasa.gov/science-news/science-at-nasa/2002/solarcells.

Liming, Drew. "Careers in Geothermal Energy." Bureau of Labor Statistics, September 2012. https://www.bls.gov/green/geothermal_energy/geothermal_energy.htm.

Locsin, Aurelio. "A Job Description for a Renewable Energy Technician." Work—Chron.com. http://work.chron.com/job-description-renewable-energy-technician-17008.html.

Maraboli, Steve. *Life, the Truth, and Being Free*. Port Washington, NY: Better Today, 2009.

Marsh, Jacob. "What Is Solar Energy?" EnergySage, January 10, 2010. https://news.energysage.com/what-is-solar-energy/.

Murphy, Jim, and Lauren Anderson, *Responsible Wind Power and Wildlife*. Issue brief from the National Wildlife Federation, January 2019. https://www.nwf.org/-/media/Documents/PDFs/NWF-Reports/2019/Responsible-Wind-Power-Wildlife.ashx.

Nelson, Gaylord. "Environment, Population, Sustainable Development: Where Do We Go from Here?" Remarks on the 25th Anniversary of Earth Day, April 22, 1995. http://www.nelsonearthday.net/nelson/quotes.php.

North American Board of Certified Energy Practitioners. "Associate Program." https://www.nabcep.org/certifications/associate-program/.

———. "NABCEP Board Certifications." https://www.nabcep.org/certifications/nabcep-board-certifications/.

———. "NABCEP Course Catalog." https://coursecatalog.nabcep.org/.

North American Electric Reliability Corporation. "System Operator Certification." https://www.nerc.com/pa/Train/SysOpCert/Pages/default.aspx.

O-Net Online. "Summary Report for 49-9099.01 Geothermal Technicians." Updated 2019. https://www.onetonline.org/link/summary/49-9099.01.

OwlGuru.com. "A Day in the Life of Solar Photovoltaic Installers." https://www.owlguru.com/day-in-life-of-solar-photovoltaic-installers/.

———. "Being a Solar PV Installer: What You Really Do." https://www.owlguru.com/career/solar-photovoltaic-installers/job-description/.

———. "Being a Solar PV Installer: What You Really Need." https://www.owlguru.com/career/solar-photovoltaic-installers/requirements/.

———. "Being a Wind Turbine Technician: What You Really Do." https://www.owlguru.com/career/wind-turbine-service-technicians/job-description/.

———. "Being a Wind Turbine Technician: What You Really Need." https://www.owlguru.com/career/wind-turbine-service-technicians/requirements/.

———. "How Much Do Wind Turbine Technicians Make in 2018." https://www.owlguru.com/career/wind-turbine-service-technicians/salary/.

Pacific Northwest Center of Excellence for Clean Energy. "Renewable Energy Technician." https://www.cleanenergyexcellence.org/careers/.

Pasadena City College. "Electrical Technology—Photovoltaic Design and Installation." *Pasadena City College 2019–2020 Catalog.* https://pasadena.edu/academics/docs/catalog19-20.pdf.

PayScale. "Average Solar Energy System Installer Hourly Pay." Updated January 4, 2020. https://www.payscale.com/research/US/Job=Solar_Energy_System_Installer/Hourly_Rate.

———. "Average Wind Turbine Technician Hourly Pay." Updated January 13, 2020. https://www.payscale.com/research/US/Job=Wind_Turbine_Technician/Hourly_Rate.

Petruzzello, Melissa. "Hydroelectric Power." *Encyclopedia Britannica.* https://www.britannica.com/science/hydroelectric-power.

Phillips, Avery. "Career Pathways and Possibilities in Renewable Energy." Renewable Energy World, February 28, 2018. https://www.renewableenergyworld.com/2018/02/28/career-pathways-and-possibilities-in-renewable-energy/.

Pierre, James. "Why Do We Need Renewable Energy Sources?" *Clean Economy Center*, November 7, 2017. https://cleaneconomycenter.org/why-do-we-need-renewable-energy-sources/.

Pulaski, Jane. "Solar Licensing Database." Interstate Renewable Energy Council, August 10, 2010. https://irecusa.org/2010/08/solar-licensing-information/.

Real de Azua, Christine. "U.S. Wind Resource Even Larger than Previously Estimated: Government Assessment." American Wind Energy Association [press release], February 18, 2010. https://web.archive.org/web/20100224133904/http://www.awea.org/newsroom/releases/02-18-10_US_Wind_Resource_Larger.html.

Richard, Michael Graham. "Wind Turbines Kill around 300,000 Birds Annually, House Cats around 3,000,000,000." *Treehugger*, September 16, 2014. https://www.treehugger.com/renewable-energy/north-america-wind-turbines-kill-around-300000-birds-annually-house-cats-around-3000000000.html.

Rouse, Margaret. "Definition: Turbine." Whatis.com. Updated January 2014. https://whatis.techtarget.com/definition/turbine.

Santa Monica College. "A Course Study for Solar Photovoltaic and Energy Efficiency," July 26, 2018. http://www.smc.edu/StudentServices/TransferServices/AreasofStudy/Documents/Associate_Degrees/solar_photovoltaic_installation_as.pdf.

Shinn, Lora. "Renewable Energy: The Clean Facts." National Resource Defence Council, June 15, 2018. https://www.nrdc.org/stories/renewable-energy-clean-facts.

Smith, Jacquelyn. "The 20 People Skills You Need to Succeed at Work." *Forbes*, November 16, 2018. https://www.forbes.com/sites/jacquelynsmith/2013/11/15/the-20-people-skills-you-need-to-succeed-at-work/#38a39de13216.

Solar Foundation. "Solar Ready Vets Network." https://www.thesolarfoundation.org/solar-ready-vets/.

Speer, Anh. "Geothermal Technician Jobs and Green Career Profile." *Renewable Energy Jobs*, December 31, 2013. http://www.isustainableearth.com/green-jobs/renewable-energy-job-profiles/geothermal-technician-jobs.

———. "Hydroelectric Plant Technician Jobs and Green Career Profile." *Renewable Energy Jobs*, January 30, 2014. http://www.isustainableearth.com/green-jobs/renewable-energy-job-profiles/hydroelectric-plant-technician-jobs.

Study.com. "HVAC Associates Degree Programs with Course Information." https://study.com/hvac_associates_degree_programs.html.

Tennessee Valley Authority. "Careers: Hydro Technician Training Program." https://www.tva.com/Careers/Hydro-Technician-Training-Program.

Trade-Schools.net, "Renewable Energy Jobs: Careers That Make a Difference in the World." https://www.trade-schools.net/articles/renewable-energy-jobs.asp#solar.

Tulsa Community College. "Engineering AS, Electrical Engineering Technology Option." *Tulsa Community College 2019–2020 College Catalog.* https://catalog.tulsacc.edu/preview_program.php?catoid=11&poid=2263.

Union of Concerned Scientists. "Achieving 100 Percent Clean Electricity," December 4, 2018. https://www.ucsusa.org/resources/achieving-100-percent-clean-electricity.

———. "Renewable Energy: Unlimited resources. Little or no pollution. Solar Power." https://www.ucsusa.org/energy/renewable-energy.

———. "How Is Electricity Measured?" July 14, 2008, updated October 22, 2013. https://www.ucsusa.org/resources/how-electricity-measured.

US Bureau of Labor Statistics *Occupational Outlook Handbook*. "How to Become a Wind Turbine Technician." https://www.bls.gov/ooh/installation-maintenance-and-repair/wind-turbine-technicians.htm#tab-4.

———. "Power Plant Operators, Distributors, and Dispatchers." https://www.bls.gov/ooh/production/power-plant-operators-distributors-and-dispatchers.htm..

———. "Wind Turbine Technicians: Job Outlook." https://www.bls.gov/ooh/installation-maintenance-and-repair/wind-turbine-technicians.htm#tab-6.

———. "Wind Turbine Technicians: Summary." https://www.bls.gov/ooh/installation-maintenance-and-repair/wind-turbine-technicians.htm.

US Department of Energy Office of Energy Efficiency & Renewable Energy. "5 of the Fastest Growing Jobs in Clean Energy," April 24, 2017. https://www.energy.gov/eere/articles/5-fastest-growing-jobs-clean-energy.

———. "Everything You Need to Know about Wind Turbine Technicians," October 23, 2017. https://www.energy.gov/eere/articles/everything-you-need-know-about-wind-turbine-technicians.

———. "How to Become a Wind Turbine Technician." https://www.bls.gov/ooh/installation-maintenance-and-repair/wind-turbine-technicians.html.

———. "Types of Hydropower Plants." https://www.energy.gov/eere/water/types-hydropower-plants.

US Energy Information Administration. "Geothermal Explained: Where Geothermal Energy Is Found." Updated December 5, 2019. https://www.eia.gov/energyexplained/geothermal/where-geothermal-energy-is-found.php.

Walters State Community College. "Clean Energy Technology." *Walters State Community College 2019–2020 Catalog and Student Handbook*. http://catalog.ws.edu/preview_program.php?catoid=19&poid=3214&returnto=1237.

Zippia. "Become a Wind Turbine Technician." https://www.zippia.com/wind-turbine-technician-jobs/.

———. "Working as a Solar Panel Installer." https://www.zippia.com/solar-panel-installer-jobs/.

About the Author

Marcia Santore is an author and artist from New England. She enjoys writing about interesting people and the fascinating things they do. She's written on many topics, including profiles of artists, scholars, scientists, and businesspeople. She has also illustrated and published several children's books. See her writing website at www.amalgamatedstory.com and her artwork at www.marciasantore.com.

EDITORIAL BOARD

Eric Evitts has been working with teens in the high school setting for twenty-three years. Most of his career has dealt with getting teens, especially at-risk students, to find and follow a career path of interest. He has developed curriculum for Frederick County Public Schools focusing on anti-bullying and career development. He is currently a counselor at South Hagerstown High School.

Danielle Irving-Johnson, MA, EdS, is currently the career services specialist at the American Counseling Association. She exercises her specialty in career counseling by providing career guidance, services, and resources designed to encourage and assist students and professionals in obtaining their educational, employment, and career goals while also promoting the importance of self-care, wellness, work-life balance and burnout prevention. Danielle has also previously served as a mental health counselor and clinical intake assessor in community agency settings assisting diverse populations with various diagnoses.

Joyce Rhine Shull, BS, MS, is an active member of the Maryland Association of Community Colleges Career Affinity Group and the Maryland Career Development Association. She presently serves as an academic adviser in higher education and teaches Professionalism in the Workplace as an adjunct professor. Her experience also includes two decades of management and career education of vocational courses and seminars for high school students.

Lisa Adams Somerlot is the president of the American College Counseling Association and also serves as director of counseling at the University of West Georgia. She has a PhD in counselor education from Auburn University and is a licensed professional counselor in Georgia and a nationally approved clinical supervisor. She is certified in Myers-Briggs Type Indicator, Strong Interest Inventory, and Strengths Quest administration.

www.ingramcontent.com/pod-product-compliance
Lightning Source LLC
Chambersburg PA
CBHW031553300426
44111CB00006BA/302